Technological Innovation in the Olive Oil Production Chain

Edited by Innocenzo Muzzalupo

Published in London, United Kingdom

IntechOpen

Supporting open minds since 2005

Technological Innovation in the Olive Oil Production Chain
http://dx.doi.org/10.5772/intechopen.78420
Edited by Innocenzo Muzzalupo

Contributors
Alev Yüksel Aydar, Alicia Penissi, Amany Basuny, Maria Lisa Clodoveo, Innocenzo Muzzalupo

First published in London, United Kingdom, 2019 by IntechOpen
IntechOpen is the global imprint of INTECHOPEN LIMITED, registered in England and Wales, registration number: 11086078, The Shard, 25th floor, 32 London Bridge Street
London, SE19SG - United Kingdom
Printed in Croatia

British Library Cataloguing-in-Publication Data
A catalogue record for this book is available from the British Library

Additional hard and PDF copies can be obtained from orders@intechopen.com

Technological Innovation in the Olive Oil Production Chain
Edited by Innocenzo Muzzalupo
p. cm.
Print ISBN 978-1-83962-645-6
Online ISBN 978-1-83962-236-6
eBook (PDF) ISBN 978-1-83962-237-3

Meet the editor

Innocenzo Muzzalupo received his degree in Biology from the University of Calabria (UNICAL) in 1993 and his PhD degree in Chemistry from the University of "La Sapienza" Rome in 1997. Currently, he is working as a researcher at the Council for Agricultural Research and Economics, Research Centre for Olive, Citrus and Tree Fruit (CREA-OFA) in Italy. After receiving his PhD degree, he was appointed as postdoctoral (1999) in Food Science at UNICAL. Between 1999 and 2008 he had a contract as Professor of Botany at UNICAL. In 2008 he was appointed as a researcher at CREA. He is an editorial board member of the *Advances in Agriculture* and *Novel Techniques in Nutrition and Food Science* journals. He has authored numerous research articles in ISI journals.

Contents

Preface

Technological innovation is a type of process that leads to the expansion and improvement of techniques and technologies for the production and distribution of products. Technological innovation can be pursued both by public institutions and private companies. Despite the name, technological innovation can also involve concepts other than technology. The best organization of the production and distribution system is also considered an innovation.

Technological innovation can be divided into:

"Product innovation" is the introduction on the market of a new product or service. It can be a radical or incremental innovation depending on whether it is a new product or the improvement of an existing product.

"Process innovation" is the introduction of a new production process that increases production efficiency, reducing production costs and/or increasing productivity. It can be a radical or incremental innovation depending on whether the innovation involves a new production process or the improvement of a pre-existing process.

Technological innovation is one of the driving factors for economic growth and business competitiveness. The introduction of technological innovation can create new markets or market segments to radically change the market balance between companies and also the same behavior habits in society.

The agri-food system, and in particular the olive oil supply chain, in Italy is of great importance, both in terms of employment and turnover. The agri-food system includes all those sectors of the economy involved in agricultural production, in the transformation of the agricultural product into a food commodity, and in distribution to the final consumer. In the upstream phase of each agri-food supply chain we can find agricultural enterprises. The agricultural enterprises have unique characteristics, which differentiate them from any other productive activity, since production is dependent on climatic and environmental factors, as well as on the biological and seasonal cycle, and the output of the productions are characterized by rapid perishability. Over the years the weight of the various actors within the agri-food system has changed. The agricultural sector is also important from economic and social points of view, because it is and has been an element for the integration of the European Union countries through the Community Agricultural Policies.

This work aims to analyze the link between the olive oil production chain and technological innovation. The global opening of markets obliges companies to make continuous efforts in research and development to innovate their products; demand is constantly changing, and therefore companies must be quick to catch these changes, trying to satisfy the needs of consumers with products that are adequate to try to remain competitive in the market.

The book highlights the aspects of technological innovation as a means for the production of high-quality olive oils, and then moves on to address topics such as the contemporary challenges of the extra virgin olive oil chain through emerging technologies and how some of these technological innovations (i.e. microwave, pulsed electric field, and ultrasound) can improve not only the quantity of the product but also, and above all, the quality. Obviously, the quality of extra virgin olive oil goes through its main antioxidant compounds, such as phenolic compounds and tocopherols, and on the effect these molecules play on human health, such as the regulation of activation of connective tissue mast cells. Finally, the sensorial aspects of virgin olive oils and their classification based on European Community regulations are not overlooked.

The purpose of this book is to provide a glimpse into the olive oil industry by presenting the thoughts of some of the scientists who are engaged in the development of new tools and ideas used to improve the quality of products, often from very different perspectives. I hope and trust that the information in this book will be used as the basis for policy and technical decisions to strengthen national efforts to conserve and utilize the treasures incorporated in the world's plant genetic resources to address the urgent problems faced by agriculture today and tomorrow.

I would like to express my deepest gratitude to all the authors who contributed to this book by sharing their valuable work. Special thanks also go to the reviewers who have used their valuable time for the improvement of the chapters. Finally, thanks to the publishing house that provided me with great professionalism in the realization of the book.

Innocenzo Muzzalupo
Research Centre for Olive,
Citrus and Tree Fruit,
Council for Agricultural Research and Economics (CREA-OFA),
Rende (CS), Italy

Section 1

Introduction

Chapter 1

Introductory Chapter: Technological Innovation as Tool for Products Qualification

Innocenzo Muzzalupo

1. Technological innovation

In a context of increasing competitive pressure, company management that is more professional in terms of marketing is an essential condition to generate perceivable value in the olive-olive oil sector.

Technological innovation has undergone unprecedented development; this evolution can offer extraordinary opportunities for product qualification which, today, have not been adequately exploited due to a lack of vision. Understanding the needs of consumers and innovations in managing the different stages of the supply chain represent the real challenge for competitiveness of olive oil companies in the near future. It represents a challenge that is destined to produce commercial proposals that are increasingly in line with the trends of current demand.

Digital innovation in the agri-food sector can guarantee competitiveness to one of the key sectors of the Italian economy, which contributes more than 11% of GDP and 9% of exports. On the one hand, Smart AgriFood[1] can reduce production costs of high-quality products, while, on the other hand, it can increase revenues thanks to greater recognition or guarantee, for example, with anti-counterfeiting or the reduction of non-compliant exported products. However, digital innovation also makes it possible to intervene to support the entire supply chain, guaranteeing sustainability to all participants in the sector, including production in the field. The data all lead in that direction; with a population of 9.7 billion and an increase in demand of 32%, according to forecasts to 2050, there is little leeway. However, digital innovation could support the entire supply chain, encouraging players in the sector, including field production [1]. However, there are still obstacles; currently less than 1% of cultivated areas are managed with these solutions.

However, for digital technologies to fully release their potential certain conditions must be fulfilled. First of all, the extension of broadband and extra-wide bandwidth to rural areas is necessary in order to ensure supply chain interconnection. Sensitivity, skill and a propensity to invest by companies, a fact not taken for granted, considering the small average size of farms, are also required [2].

---

[1] The term Smart AgriFood identifies, in a nutshell, a vision of the future of the agricultural and agri-food supply chain according to which, thanks to digital technologies, the entire sector will increase its competitiveness.

It is necessary to remember that the industry 4.0[2] plan aims to have all Italian companies connected at 30 Mbps by 2020 and 50% connected at 100 Mbps by 2050 [3]. On the pulverization of our agricultural system (a long and fractioned supply chain with a proliferation of small companies), a single figure is sufficient: the average size is 12 ha (compared to 58 ha in France), with 2.5 employees which constitutes an undoubted brake on investments, even in managerial resources. It is essential that Italy focus on both quality and digital technology to compete, while the few large companies could also focus on robotics. An important figure, that has emerged from research and which should guide choices, is that 80% of consumers use smartphones for purchasing decisions.

When should planting and irrigation take place? When and how much fertilizer? When should action be taken to prevent pathologies? How can yields be maximized based on different land performance? A cross-analysis of environmental, climatic and cultural factors allows the execution of targeted actions, to improve crop yields, use less pesticides and affect product quality. In order to benefit from these opportunities, which save both time and money, it is necessary to switch from precision farming (GPS and satellite-driven tractors), which have been discussed since the 1990s, to interconnected farming, the so-called Internet of Farming, with its new instruments (drones, sensors, the Internet of Things and Big Data). A survey was conducted for 220 solutions offered in Italy by more than 70 companies: 89% support precision farming while only 11% have enabled the Internet of Farming. In particular, the majority exploits data and analytics (73%), 41% uses the Internet of Things and 57% uses processing systems and user interface software.

Data management is the key element in agriculture 4.0,[3] but it must be translated into information and be transferred to all operators in the supply chain. One of the key issues is the capacity of reading, harmonizing and standardizing data given that they come from different sources; this is the reason why researchers have warned the need to invest in training as well as to overcome obstacles to innovation. Just tractors in Italy generate over 1 million gigabytes in a year, in addition to environmental, warehouse, farm and more general data of a corporate nature; however, today this information is scarcely valued [4].

Today the perception of quality in a product has changed: taste is no longer sufficient; the complex form is a heptagon where, on each side, there are factors that influence it, such as food safety, nutrition, provenance of raw materials, social and environmental impacts of production processes (such as animal welfare), appearance, taste and aroma, and service. Based on an analysis conducted by researchers in this sector, in 57 case studies, digital innovation allows companies to improve the

[2] Industry 4.0 is a name given to the current trend of automation and data exchange in manufacturing technologies. It includes cyber-physical systems, the Internet of things, cloud computing and cognitive computing. Industry 4.0 is commonly referred to as the fourth industrial revolution. Industry 4.0 fosters what has been called a "smart factory". Within modular structured smart factories, cyber-physical systems monitor physical processes, create a virtual copy of the physical world and make decentralized decisions. Over the Internet of Things, cyber-physical systems communicate and cooperate with each other and with humans in real-time both internally and across organizational services offered and used by participants of the value chain.

[3] Agriculture 4.0 is therefore for cooperation and sharing in terms of data and information, between different machines, between different operators, along the entire supply chain. You look at the whole process, no longer at the single step: the tractor communicates with the harvester and perhaps sends a signal to the local supplier, for example when a component does not work. The agricultural entrepreneur can have a complete view of all the activities of the machines in the field and make sure that they are working at their best; can monitor the results and costs of operations, at every stage of the process, putting them in relation to the possible final price of its product on the market. This is, at least in theory, Agriculture 4.0.

qualities on all the dimensions of the heptagon. A total of 51% of the companies uses digital technologies to valorize qualities of origin, especially in the case of products with a high added value (such as wine, cocoa and coffee); 46% used digital innovation to improve food safety and 25% concentrated on production methods, above all for aspects linked to environmental impact, animal welfare and agri-food traditions of different territories; lastly, in 12% of the cases, the businesses used technology to improve service quality, adopting innovative solutions to communicate product information (nutritional advice) and process information (origins, traceability and environmental impact) to consumers [5].

It took a disaster such as 'mad cow' disease to accelerate the traceability plan; today, 36% of the analyzed agri-food companies, thanks to digital solutions, achieved a reduction in the times and costs connected with harvest processes, data management and transmission [6]. Digital solutions permit interventions aimed at food safety along the food chain thus avoiding financial damage. But they can also be used to combat counterfeiting to protect the protected designation of origin (PDO) and protected geographical indication (PGI) systems, for greater information for the consumer. The sectors most involved in technological innovation for traceability are the fruit and vegetable sector (30%), the meat supply chain (23%), dairy products (14%) and coffee-cocoa (12%). The instruments used to improve traceability are barcodes (39%), RFIds (Radio-Frequency Identification, 32%) management systems (32%), Big Data (30%) and mobile technologies (21%), while innovative technologies such as the IoT and blockchains have yet to be explored [7].

From 2011 until today, 481 international start-ups have been created, of which 12% are Italian. Of these start-ups, 218 regard e-commerce and the area with the greatest presence is the United States. The most important sectors include the fruit and vegetable sector, with 17% of international start-ups. Precision agriculture and food quality are the most explored and most interesting application areas for investors. Also in Italy, the most important sector is the fruit and vegetable sector (14% of Italian start-ups), followed by wine (9%) and cereal (7%). Quality and environmental sustainability are the areas in which they are most active, with 50% of the funds raised, followed by precision agriculture (35%) and food quality (29%) [8].

How is the health of Italian extra virgin olive oil, one of the flags of the Made in Italy brand? One might say it has its ups and downs; quality oil in circulation, for which demand is increasing, constitutes less than half, 39.2% to be precise. This has been confirmed by a study conducted by Symbola (Foundation for Italian Qualities), CREA (Council for Agricultural Research and Analysis of Agricultural Economics) in collaboration with Coldiretti and Unaprol (National Union of Olive Producers Associations). The report (called Poq, quality internal product) was presented in the Coldiretti pavilion at Milan Expo; it evaluated not only the organoleptic qualities, but also the entire chain including 102 indicators, the soil, grinding, respect for the environment, the use of pesticides, water consumption, human capital management as well as distribution.

The results say that a segment of producers invests in quality while about 60% do not. Yet, that is the winning path. Even in times of crisis, in contrast with the overall trend of the sector, in Italy, it is precisely the consumption of quality that continues to record growth. It would therefore need a turning point, the same that in the past affected the wine sector, where, thanks also to the methanol wine scandal (23 deaths), quantity has been abandoned for quality. Today, numbers justify that change with a value that has risen by six, seven times, despite the fact that production has fallen by 50%. The Piq, therefore, may prove to be a useful database not only for institutions, but also for consumers, who are often poorly informed and do not understand what lies behind such marked price differences, from €2.50 to over €9 for a liter bottle of olive oil [8].

Italian oil, with a turnover of €3 billion, has important numbers. It would be enough to say that our country with 43 products holds the record of oils certified by the European brand (PDO or PGI) [7].

Italy is the second largest global producer of olive oil and constitutes a fifth of European production. It is, furthermore, the third producer of table olives and, on the topic of biodiversity, it is unbeatable due to 538 diverse cultivar types [9]. Also in terms of label transparency, Italy is second to none, having the largest network of olive traceability and monitoring on a European level [10].

Yet, in recent years, negative signals have appeared: in 2014, the national harvest fell by 35% due to a 38% increase in imported oil. The year 2015 recorded the peak of imports of foreign olive oil, in particular of Tunisian olive oil which increased by 681% in the first trimester of the year. Poor quality oil is often at risk of counterfeiting and alteration. Seizures by NAS (Nuclei Anti-Sophysation and health of the Italian Army) grew by 483% in 7 years. Extra virgin olive oil (which must be obtained exclusively with mechanical models and whose acidity must not exceed 0.8%) has been found to have altered with the addition of refined oils and those extracted with solvents, or poor-quality oils, such as hazelnut [11]. A useful step in this direction was the entrance into force of a community regulation, last November, which made the anti-topping cap for the extra virgin olive oil and the virgin compulsory in restaurants and bars. This measure seeks to prevent the empty bottle being refilled, which carries a fine from €1000 to €8000 euros to be paid by the operator.

A positive step forward comes from the go-ahead by the House to the agriculture legal decree which allocates €32 million for the national olive oil plan with important structural measures for the Italian supply chain.

Finally, from a nutritional point of view, it should be remembered that extra virgin olive oil is one of the components of the Mediterranean diet, registered by UNESCO as world heritage. According to the tables of the CREA Research Centre for Food and Nutrition, an extra virgin olive oil has only about 14% saturated fats (whose intake levels should be kept low). Just to give an idea, coconut oil has 86% and the infamous palm oil 47% [12].

Acknowledgements

This work was supported by grants from the POR Calabria FESR-ESF 2014-2020 Action 1.2.2, S3 strategies. “Extra virgin olive oil digital ID management - ODIN” Project.

Conflict of interest

The author declares no conflict of interest.

Author details

Innocenzo Muzzalupo
Centro di Ricerca Olivicoltura, Frutticoltura e Agrumicoltura, Consiglio per la Ricerca in Agricoltura e L'analisi Dell'economia Agraria (CREA-OFA), CS, Italy

*Address all correspondence to: innocenzo.muzzalupo@crea.gov.it

References

[1] Kagermann H, Wahlster W, Helbig J, editors. Recommendations for Implementing the Strategic Initiative Industrie 4.0: Final Report of the Industries 4.0 Working Group; 2013

[2] Lasi H, Kemper HG, Fettke P, Feld T, Hoffmann M. Industry 4.0. Business & Information Systems Engineering. 2017;**4**(6):239-242

[3] Egon M, Xiao-Li C, Ralph R. Challenges and requirements for the application of industry 4.0: A special insight with the usage of cyber-physical system. Chinese Journal of Mechanical Engineering. 2017;**30**(5):1050-1057

[4] Ministry of the Policies Agricultural, Food, Forestry and Tourism. Report sullo stato dell'arte dell'Agricoltura di Precisione in Italia. 2015. Available from: www.politicheagricole.it

[5] Gatto S. Applicazione delle Tecnologie di Agricoltura di Precisione nella Coltivazione del Mais in una Azienda Cerealicola-Zootecnica. 2012. Available from: http://tesi.cab.unipd.it/42408/

[6] Eurostat. Statistiche Sulla Struttura Delle Aziende Agricole. 2015. Available from: https://ec.europa.eu/eurostat/statistics-explained/index.php?title=Farm_structure_statistics/it&oldid=370120

[7] European Union Commission Regulation. (EEC) No 2568/91. Brussels: Official European Commission Journal; 1991

[8] International Olive Oil Council. World Olive Oil Balances for 2017/18 Market Newsletter; IOOC. 2018. Available from: ///C:/Users/alev/Downloads/CONSOMMATION1_ANG.pdf

[9] Muzzalupo I. Olive Germplasm—Italian Catalogue of Olive Varieties. Rijeka, Crotia: InTech; 2012. Available from: https://www.intechopen.com/books/olive-germplasm-italian-catalogue-of-olive-varieties/olive-germplasm-italian-catalogue-of-olive-varieties

[10] Chiappetta A, Muto A, Muzzalupo R, Muzzalupo I. New rapid procedure for genetic characterization of Italian wild olive (Olea europaea) and traceability of virgin olive oils by means of SSR markers. Scientia Horticulturae. 2017;**226**(19):42-49

[11] Muzzalupo I. Breaking news on the authenticity of olive oils by means of molecular markers. Novel Techniques in Nutrition and Food Science. 2017;**1**(1):000505. Available from: https://crimsonpublishers.com/ntnf/pdf/NTNF.000505.pdf

[12] Council for Agricultural Research and Analysis of Agricultural Economics. 2018. Available from: https://www.crea.gov.it/it

Section 2

Novel Technologies

Chapter 2

Emerging Extraction Technologies in Olive Oil Production

Alev Yüksel Aydar

Abstract

In the field of olive oil extraction, current scientific research has focused on improving quality, paying particular attention to optimizing the efficiency of extraction and reducing the duration of the process. Recently, studies have been conducted to improve the traditional malaxation process and obtain positive effects on both oil production and consumption. With these aims, emerging technologies including microwave (MW), pulsed electric field (PEF), and ultrasound (US) have been applied to conventional virgin olive oil extraction process. In this chapter, most recent studies that focused on adaptation of emerging technologies to traditional extraction to increase the yield of olive oil or some minor compounds and bioactive components present in olive oil including tocopherols, chlorophyll, carotenoids, and phenolic compounds have been compiled.

Keywords: novel technologies, olive oil, extraction, ultrasound, microwave

1. Introduction

The conventional extra virgin olive oil (EVOO) extraction method consists of three main processes, which are crushing, malaxation, and centrifugation [1]. After washing olive fruits, they are crushed using a stone-mill, hammers, disc crushers, de-stoning machines, or blades [2]. The purpose of this step is to facilitate the release of the oil droplets from the Elaioplasts. The minimum size for the continuous separation process of olive oil is 30 μm, but only 45% of the oil droplets have a diameter greater than 30 μm after crushing increases. This ratio reaches 80% with the formation of larger diameter drops from the oil droplets by malaxation [3]. Malaxation and crushing are main steps that affect the quality and yield of oil [4]. A flow chart of extra virgin olive oil extraction is shown in **Figure 1**.

Conventional techniques in olive oil extraction have not changed significantly for last 20 years [5]. However, in line with research findings and new techniques developed by market demand, the ongoing food industry has become very active in looking for new methods for food innovation. But, it is still very uncommon for the food industry to develop and adopt advanced processing techniques in the direction of consumers' increasing food safety and quality requirements [6]. Researchers working on the development of food technology are making great efforts to develop and implement "minimal processing" strategies to remove the negative effects of traditional food processing methods. The most general definition of minimal processing can be: preserving the nutritional quality and sensory qualities of food by heat application, which is the basic protection step in food processing, for a shorter

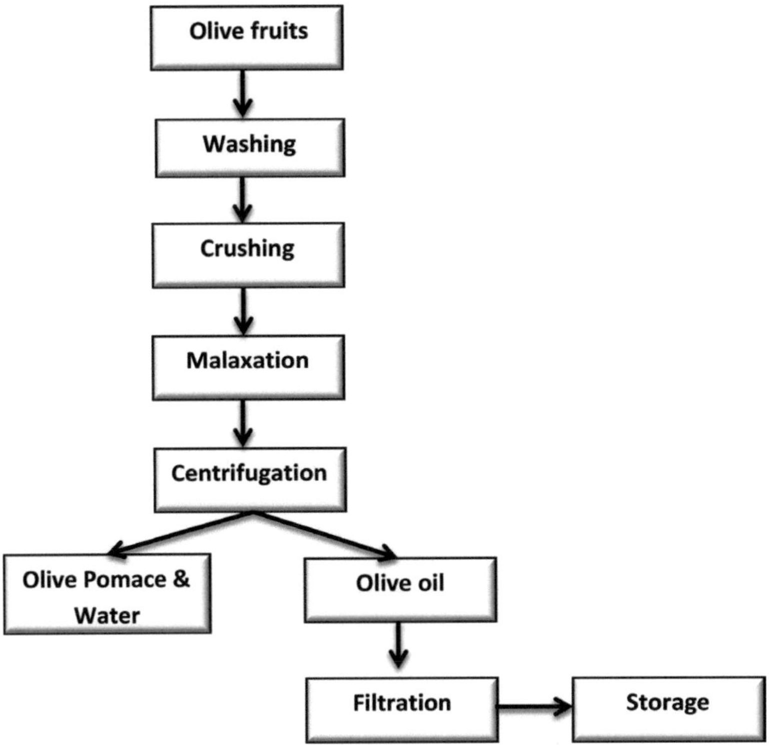

Figure 1.
Flow chart of olive oil extraction.

period of time. [7]. Emerging technologies including microwave, high-pressure processing, pulsed light, radio frequency, Ohmic heating, ultrasound, and pulsed electric field (PEF) are widely applied emerging minimal processes in the food industry.

In recent years, novel technologies such as ultrasound, pulsed electric field, or microwave have been adopted in olive oil extraction [1, 8–10] because of their positive effects including enhanced extraction efficiency, reduced extraction time, increased yield, and low energy consumption.

Ultrasound is one of the main emerging technologies widely used in various extraction processes of plant materials [11, 12]. In order to enhance oil extraction, ultrasound can be applied to the olive paste due to its mechanic effect on the cell membranes, which induces them to release oil easily from vacuoles with a considerably lower malaxation time and higher oil quality and yield [2, 5, 10, 13–18]. In addition to the extraction process, ultrasound was also investigated in numerous studies on food processing methods including emulsification, filtration, crystallization, inactivation of enzymes and microorganisms, thawing, and freezing on foods [19, 20].

It has been demonstrated that pulsed electric field (PEF), another non-thermal technology, is effective for reversible or irreversible permeabilization of cell membranes in several plant tissues, without significant temperature increase [8]. PEF technology, which has been used in the field of food science since 1960, is based on the principle of exposing liquid or solid food products to an electric field causing pores in cell membranes [6].

Microwave-assisted extraction (MAE) is an alternative oil extraction method in recent years. Since microwave provides more rapid heating and destruction of biological cell structures in a shorter time, it is a more efficient extraction method than conventional processes. Other important advantages of this method are obtaining high-quality oil and low energy requirement, which cause a significant reduction in environmental impact and financial costs [21].

More emphasis has been placed on the understanding of a superior EVOO quality based on the preservation of the sensory characteristics and positive health properties of olive oil in recent years. This aspect of EVOO quality is strongly related to the presence of phenolic and volatile compounds [13, 22]. Therefore, utilization of an emerging technology in olive oil extraction should not only increase oil yield, but also protect and improve the bioactive oil compounds and the oil quality. Recent studies that applied emerging technologies to olive oil extraction are summarized in **Table 1**.

Variety of olive	Emerging technology*	Investigated parameters	Dependent variables	Refs.
Edremit	HPU	Ultrasound time, ultrasound temperature, malaxation time	Oil yield, acidity, peroxide value, and antioxidant properties	[1]
Coratina	HPU	Ultrasound application step (After crushing/ before crushing)	Olive paste temperature, energy balance, oil yield, quality indices of oil, minor compounds	[10]
Picual	HPU	Direct/indirect application of ultrasound	Olive paste temperature	[15]
Picual	HPU	Continuous ultrasound application before centrifugation	Oil yield, quality indices, volatile and minor compounds, fatty acid composition	[23]
Edremit, Gemlik, Uslu	HPU	Ultrasound and malaxation time	Oil yield, UV absorbance values acidity, peroxide value, total phenolic content	[24]
Picual	HPU	Olive paste flow, HPU intensity, fruit temperature, olive moisture, and fat content	Olive paste temperature	[25]
Ogliarola Barese	HPU, MW	Thermal effect of US and MW	Malaxation time, oil yield, quality characteristics, and energy efficiency	[14]
Arbequina	PEF	Malaxation time and temperature, electric field strength (kV/cm)	Oil yield, acidity, quality characteristics, total and individual phenols	[8]
Arroniz	PEF	Application of PEF	Oil yield, acidity, quality characteristics, total phenols, sensory properties	[9]
Unspecified/Olive pomace used	MW	Microwave power, irradiation time, solvent-to-sample ratio	Oil yield, physicochemical oil properties	[26]

Variety of olive	Emerging technology*	Investigated parameters	Dependent variables	Refs.
Chemlal	MW	The extraction time, acetic acid content in hexane, irradiation power	Oil yield, total phenols, quality parameters	[27]
Unspecified/Olive oil used	MW	Microwave heating times	Quality and physicochemical properties, oil color	[28]
Coratina	MW/MS	MW and MS combined effect	Rheological properties, oil yield	[29]
Peranzana	MW	Malaxation time and MW	Energy consumption, oil yield, structure modifications of olive pastes	[30]
Coratina	HPU	Sonication time	Oil yield, oil quality indices, phenolic composition	[31]

**HPU: High-power ultrasound, PEF: Pulsed electric field, MS: Megasonic treatment, MW: Microwave.*

Table 1.
Emerging extraction technologies used in olive oil production.

2. Ultrasound applications in olive oil extraction

In the olive oil industry, ultrasound is the one of the most promising technologies because of its powerful mechanical and mild thermal effects [32]. Many researchers have used this technology to investigate its effects on overall olive oil quality and yield in the last decade [1, 10, 14–16, 23–25, 33]. In recent years, it has been discovered that using a stronger ultrasound (>1 W/cm^2) at a lower frequency (generally around 20–50 kHz), which is also called high-power ultrasound (HPU) (usually around 20–50 kHz), is physically effective in altering the properties of a substance or inactivating microorganisms [6, 7].

High-power ultrasound application in olive oil extraction was first performed by Jiménez et al. [15] under discontinuous conditions. In their studies investigating the effects of direct and indirect ultrasound, they found that direct sonication provided better extractability in high-moisture olives (>50%) while greater extractability was obtained by indirect sonication in low-moisture olive fruits (<50%) [15].

Enrichment of olive oil with main phenols in olive leaves using ultrasound has been studied by researchers [34, 35]. Achat et al. [34] used ultrasound to enrich olive oil with oleuropein both on a laboratory and a pilot plant scale. The ultrasound-assisted extraction method greatly facilitated the enrichment of VOO in phenolic compounds compared to conventional processes. They found that tyrosol and hydroxytyrosol, main phenolic compounds present in olive oil, were not significantly degraded by sonication [34].

Clodoveo et al. [10] investigated ultrasound application on olive fruits submerged in a water bath before crushing and also on olive paste after crushing. The purpose of their study was to test the possibility of decreasing the malaxation time. Reduction in the malaxation time and improvement in oil yields and its minor nutritional compounds were attained by ultrasound technology. The results were better in oils obtained by sonication of olives in water bath than those obtained by sonication of olive paste [10].

Bejaoui et al. [25] applied HPU to olive paste through the pipe before centrifugation with continuous conditions. They observed that when the oils were extracted without ultrasound, the extraction yield was 46.83% ± 0.83, while ultrasound treatment of olive paste produced a significant increase in extraction yields to 52.75% ± 1.39.

Aydar et al. [1] used an ultrasound bath in olive oil extraction to find optimum ultrasound-assisted olive oil extraction conditions based on maximum oil yield and minimum free acidity. The acidity of the oils for all experiments was below the legal limit (<8 g oleic acid/kg oil) established for the category of EVOO [36]. The most important impact on the extraction yield and the acidity ($p < 0.05$) was due to the malaxation temperature. They also observed that ultrasound time had no significant effect ($p > 0.05$) on the acidity and yield [36].

The effect of malaxation time combined with the use of ultrasound on the oil yield, oxidative and quality characteristics of EVOOs extracted from different Turkish olive cultivars was studied by Aydar [24]. It was found that different sonication and malaxation time combinations did not cause difference ($p > 0.05$) in the Edremit oil yield and extractability indexes, while they were significantly different in Uslu and Gemlik oils. In that study, oils obtained by 8 min of ultrasound application and 22 min of malaxation had highest oil yield and chlorophyll and carotenoid contents. [24].

3. PEF applications in olive oil extraction

Olive paste was exposed to PEF technology involving 50 monopolar pulses of 3 μs at an electric field strength of 1 kV/cm (1.47 kJ/kg) and 2 kV/cm (5.22 kJ/kg) and a frequency of 125 Hz. PEF did not result in any significant differences in fatty acid composition and sensorial properties of oil. In sensorial properties point, panelists evaluated the oil subjected to PEF was less bitter and pungent, and more fruity than the untreated oils. The PEF treatment was very effective to increase the oil yield when combined with malaxation. The oil yield as was high as 14.10% when the olive paste was subjected to PEF at 2 kV/cm and malaxated for 30 min at 15°C. However, the extraction yield was reduced by 50% when no malaxation was applied to olive paste compared to those malaxated for 30 min. [8]

Effect of the use of pulsed electric field (PEF) technology on Arroniz olive oil production in terms of extraction yield and chemical and sensory quality has been evaluated by Puértolas and Marañón [9]. Extraction yield increased by 13.3% in PEF-treated samples (2 kV/cm, 11.25 kJ/kg) compared to control. In addition, the total phenolic content, total phytosterol, and total tocopherol of olive oil extracted with PEF showed significantly higher values (11.5, 9.9, and 15.0%, respectively) than the control group. [9]

4. Microwave applications in olive oil extraction

Over the last few decades, microwave treatments in food processing have gained popularity because of their low heat treatment times, operational simplicity, and high heating rates, which result in lower maintenance requirements. The microwaves obtained from household ovens and many industrial applications are produced efficiently by permanent wave magnetrons (**Figure 2**) [6].

The effect of heating with microwave and its comparison with conventional heating and ultrasound heating on crushed olives was investigated by Clodoveo and Hbaieb [14]. Results showed that the main quality parameters legally established (acidity, peroxide value, and specific extinction coefficients (K232 and K270)) to evaluate VOO were not affected by the microwave and ultrasound treatments. Moreover, the malaxation time was decreased and extraction yield was improved by

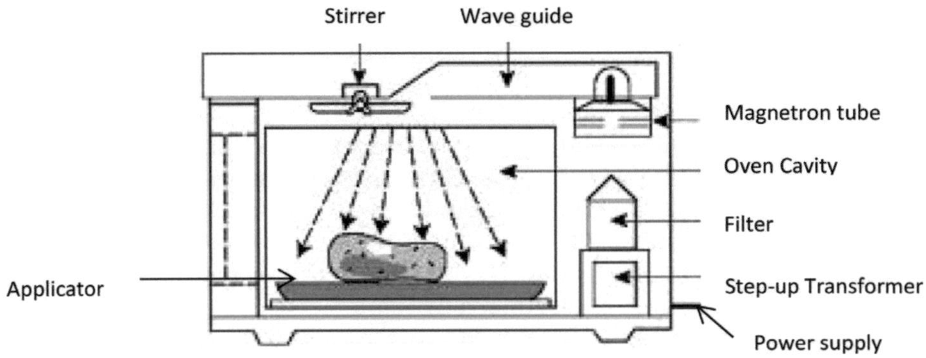

Figure 2.
Microwave oven parts.

ultrasound and microwave treatments compared with the oils that were extracted from the olive paste without malaxation. [14].

Yanık et al. investigated microwave-assisted solvent extraction (MASE) parameters on olive pomace oil. The yield of oil obtained by conventional extraction was lower than that of oil obtained by microwave extraction from olive pomace. It demonstrated that microwave-extracted oils had higher total phenolic (985 mg caffeic acid/kg oil) and tocopherol compounds (278.07 mg/kg oil), also lower peroxide value (17.8 meq O_2/kg oil) and polycyclic aromatic hydrocarbons (PAH) (0.44 µg benzo(α) pyrene/kg) compared to oils extracted by conventional industrial methods. [26]

The effect of microwave-assisted solvent extraction at two different radiation power values (170 and 510 W) combined with acetic acid on yield and physicochemical properties of olive oil was studied by Kadi et al. [27]. The UV absorbance values were highest in oils treated with 510-W microwave and 7.5% acetic acid content. Since microwave radiations accelerate the disruption of cells and oil release, they observed similar results to those of previous researchers who also achieved better oil extractability [27].

Malheiro et al. determined the effect of different microwave heating times (1, 3, 5, 10, and 15 min) on three Portuguese olive oils of different origins, one from the north, "Azeite de Trás-os-Montes" protected designation of origin (PDO); one from the center, "Azeites da Beira Interior" PDO; and one from the south of Portugal, "Azeite de Moura". They evaluated the effect of MW time on free acidity; peroxide value (PV); specific extinction coefficients (K232 and K270); color; and chlorophyll, carotenoid, and tocopherol content of oils. The carotenoids and chlorophyll pigments, which are also significant in determining olive oil stability, decreased by microwave treatment [28].

Leone et al. [30] determined the effect of microwave treatment on oil yield, structure modifications of olive pastes, and total energy consumption for a whole extraction process. The oil extractability was not significantly different from traditional extraction; however, the electrical power consumption using a microwave prototype system was higher by 24% [30].

The possibility of combining megasonic and microwave treatment in a continuous olive oil extraction system to enhance olive oil extractability was examined by Leone et al. [29]. The utilization of combined megasonic and microwave treatment to olive paste resulted in a consistent reduction of viscosity. In result, both microwave and megasonic technologies have improved the oil extractability performance by lowering the consistency of the olive paste [29].

In recent years, infrared spectroscopy, computer vision, machine olfaction technology, electronic tongues, and dielectric spectroscopy are some of the main sensing technologies applied to the virgin olive oil production process. Infrared spectroscopy can also be used to evaluate the official quality parameters of olive fruits and oil [37].

5. Conclusions

Worldwide, the total consumption of olive oil increased from 1,666,500 tons in 1990/1991 to 2,978,000 tons in the period of 2017/2018 after 27 years [30]. Recent studies on emerging extraction techniques aim to improve the quality and physicochemical properties of oils and reduce the processing time and energy consumed during extraction compared to traditional methods. Ultrasound, microwave, and pulsed electric field technologies have been successfully applied to olive oil extraction, and several positive impacts on oil yield and quality have been observed. Results show combining these emerging technologies could assist in the development of a continuous olive oil extraction process with a higher extractability than the traditional batch process without significant decrease in oil quality. Long-term stability and sensory studies should also be done to evaluate the long-term effects of these new technologies and to ensure their advantages.

Conflict of interest

The author declares that she has no "conflict of interest."

Author details

Alev Yüksel Aydar
Department of Food Engineering, Faculty of Engineering, Manisa Celal Bayar University, Manisa, Turkey

*Address all correspondence to: alevyuksel.aydar@cbu.edu.tr

References

[1] Aydar AY, Bagdatlioglu N, Köseoglu O. Effect of ultrasound on olive oil extraction and optimization of ultrasound-assisted extraction of extra virgin olive oil by response surface methodology (RSM). Grasas y Aceites. 2017;**68**:1-11 [Epub ahead of print]. DOI: 10.3989/gya.1057162

[2] Veneziani G, Sordini B, Taticchi A, Esposto S, Selvaggini R, Urbani S, Di Maio I, Servili M. Improvement of Olive Oil Mechanical Extraction: New Technologies, Process Efficiency, and Extra Virgin Olive Oil Quality. Products from Olive Tree Dimitrios Boskou and Maria Lisa Clodoveo. IntechOpen. 2016. DOI: 10.5772/64796. Available from: https://www.intechopen.com/books/products-from-olive-tree/improvement-of-olive-oil-mechanical-extraction-new-technologies-process-efficiency-and-extra-virgin

[3] Boskou D. Olive oil, Chemistry and Technology. Thessaloniki, Greece: AOCS; 2006. [Epub ahead of print]. DOI: 10.1159/000097916

[4] Clodoveo ML, Hbaieb RH, Kotti F, et al. Mechanical strategies to increase nutritional and sensory quality of virgin olive oil by modulating the endogenous enzyme activities. Comprehensive Reviews in Food Science and Food Safety. 2014;**13**:135-154

[5] Clodoveo ML. An overview of emerging techniques in virgin olive oil extraction process: Strategies in the development of innovative plants. Journal of Agricultural Engineering. 2013;**44**:49-59 [Epub ahead of print]. DOI: 10.4081/jae.2013.s2.e60

[6] Sun D-W. Emerging Technologies for Food Processing. 2nd ed. Dublin: Elsevier Inc.; 2014

[7] Baysal T, İçier F. Gıda Mühendİslİğİnde Isıl Olmayan Teknolojİler. Bornova, İzmir: Nobel yayıncılık; 2012

[8] Abenoza M, Benito M, Saldaña G, et al. Effects of pulsed electric field on yield extraction and quality of olive oil. Food and Bioprocess Technology. 2013;**6**:1367-1373

[9] Puértolas E, Martínez de Marañón I. Olive oil pilot-production assisted by pulsed electric field: Impact on extraction yield, chemical parameters and sensory properties. Food Chemistry. 2015;**167**:497-502

[10] Clodoveo ML, Durante V, La Notte D. Working towards the development of innovative ultrasound equipment for the extraction of virgin olive oil. Ultrasonics Sonochemistry. 2013;**20**:1261-1270

[11] Aydar AY. Utilization of Response Surface Methodology in Optimization of Extraction of Plant Materials. United Kingdom: Intech Open; 2018. pp. 157-169

[12] Amirante R, Distaso E, Tamburrano P, et al. Acoustic cavitation by means ultrasounds in the extra virgin olive oil extraction process. Energy Procedia. 2017;**126**:82-90

[13] Clodoveo ML. New advances in the development of innovative virgin olive oil extraction plants: Looking back to see the future. Food Research International. 2013;**54**:726-729

[14] Clodoveo ML, Hachicha Hbaieb R. Beyond the traditional virgin olive oil extraction systems: Searching innovative and sustainable plant engineering solutions. Food Research International. 2013;**54**:1926-1933

[15] Jiménez A, Beltrán G, Uceda M. High-power ultrasound in olive paste pretreatment. Effect on process yield and virgin olive oil characteristics. Ultrasonics Sonochemistry. 2007;**14**:725-731

[16] Clodoveo ML, Durante V, La Notte D, et al. Ultrasound-assisted extraction of virgin olive oil to improve the process efficiency. European Journal of Lipid Science and Technology. 2013;**115**:1062-1069

[17] Clodoveo ML, Camposeo S, Amirante R, et al. Research and Innovative Approaches to Obtain Virgin Olive Oils with a Higher Level of Bioactive Constituents. AOCS Press. 2015. [Epub ahead of print]. DOI: 10.1016/ B978-1-63067-041-2.50013-6

[18] Clodoveo ML, Dipalmo T, Schiano C, et al. What's now, what's new and what's next in virgin olive oil elaboration systems? A perspective on current knowledge and future trends. Journal of Agricultural Engineering. 2014;**45**:49

[19] Bermúdez-aguirre D, Mobbs T, Barbosa-cánovas GV. Ultrasound Technologies for Food and Bioprocessing. Springer. 2011. [Epub ahead of print]. DOI: 10.1007/978-1-4419-7472-3

[20] Chemat F, Zill-E-Huma R, Khan MK. Applications of ultrasound in food technology: Processing, preservation and extraction. Ultrasonics Sonochemistry. 2011;**18**:813-835

[21] Çavdar HK, Yanık DK, Gök U, et al. Optimisation of microwave-assisted extraction of pomegranate (*Punica granatum* L.) seed oil and evaluation of Its physicochemical and bioactive properties. Food Technology and Biotechnology. 2017;**55**:86-94

[22] Taticchi A, Esposto S, Veneziani G, et al. The influence of the malaxation temperature on the activity of polyphenoloxidase and peroxidase and on the phenolic composition of virgin olive oil. Food Chemistry. 2013;**136**:975-983

[23] Bejaoui MA, Beltrán G, Sánchez-Ortiz A, et al. Continuous high power ultrasound treatment before malaxation, a laboratory scale approach: Effect on virgin olive oil quality criteria and yield. European Journal of Lipid Science and Technology. 2016;**118**:332-336. [Epub ahead of print]. DOI: 10.1002/ejlt.201500020

[24] Aydar AY. Physicochemical characteristics of extra virgin olive oils obtained by ultrasound assisted extraction from different olive cultivars. International Journal of Scientific and Technology Research. 2018;**4**:1-10

[25] Bejaoui MA, Beltran G, Aguilera MP, et al. Continuous conditioning of olive paste by high power ultrasounds: Response surface methodology to predict temperature and its effect on oil yield and virgin olive oil characteristics. LWT - Food Science and Technology. 2016;**69**:175-184

[26] Yanık DK. Alternative to traditional olive pomace oil extraction systems: Microwave-assisted solvent extraction of oil from wet olive pomace. LWT - Food Science and Technology. 2017;**77**:45-51

[27] Kadi H, Moussaoui R, Djadoun S, et al. Microwave assisted extraction of olive oil pomace by acidic hexane. Iranian journal of chemistry and chemical engineering. 2016;**35**:73-79

[28] Malheiro R, Oliveira I, Vilas-Boas M, et al. Effect of microwave heating with different exposure times on physical and chemical parameters of olive oil. Food and Chemical Toxicology. 2009;**47**:92-97

[29] Leone A, Romaniello R, Tamborrino A, et al. Microwave and megasonics combined technology for a continuous olive oil process with enhanced extractability. Innovative Food Science and Emerging Technologies. 2017;**42**:56-63

[30] Leone A, Tamborrino A, Zagaria R, et al. Plant innovation in the olive

oil extraction process: A comparison of efficiency and energy consumption between microwave treatment and traditional malaxation of olive pastes. Journal of Food Engineering. 2015;**146**:44-52

[31] Clodoveo ML, Paduano A, Di Palmo T, et al. Engineering design and prototype development of a full scale ultrasound system for virgin olive oil by means of numerical and experimental analysis. Ultrasonics Sonochemistry. 2017;**37**:169-181

[32] Amirante R, Paduano A. Ultrasound in Olive Oil Extraction. United Kingdom: Intech Open; 2018. pp. 43-53

[33] Amirante R, Clodoveo ML. Developments in the design and construction of continuous full-scale ultrasonic devices for the EVOO industry. European Journal of Lipid Science and Technology. 2017;**119**:1-5

[34] Achat S, Tomao V, Madani K, et al. Direct enrichment of olive oil in oleuropein by ultrasound-assisted maceration at laboratory and pilot plant scale. Ultrasonics Sonochemistry. 2012;**19**:777-786

[35] Japón-Luján R, Janeiro P, De Castro MDL. Solid-liquid transfer of biophenols from olive leaves for the enrichment of edible oils by a dynamic ultrasound-assisted approach. Journal of Agricultural and Food Chemistry. 2008;**56**:7231-7235

[36] European Union Commission Regulation. (EEC) No 2568/91. Brussels: Official European Commission Journal; 1991

[37] Beltrán Ortega J, Martínez Gila DM, Aguilera Puerto D, et al. Novel technologies for monitoring the in-line quality of virgin olive oil during manufacturing and storage. Journal of the Science of Food and Agriculture. 2016;**96**:4644-4662

Chapter 3

Does the Introduction of Ultrasound in Extra-Virgin Olive Oil Extraction Process Improve the Income of the Olive Millers? The First Technology for the Simultaneous Increment of Yield and Quality of the Product

Maria Lisa Clodoveo, Filomena Corbo and Riccardo Amirante

Abstract

Olive oil is an important product of the European agro-alimentary sector. The current olive oil extraction process can be further improved in order to overcome the weaknesses of the actual system in terms of non-continuity, reduction of oil in waste, sustainability, and improvement of quality both in the healthy and sensory perspective. Many innovative approaches have been developed to improve the olive oil extraction process. However, not all the proposed innovations have the opportunity to effectively reach a technological level of readiness close to "ready for the market." An innovator should simultaneously evaluate the aptitude of its invention to turn into a widely used commercial product both under the technological and the marketing perspectives. Under the technological point of view, an innovation should be effective, so, adequate to accomplish a purpose, and efficient, so, able to perform or functioning in the best possible manner with the least waste of time and effort. Under the marketing point of view, an innovation should be able to develop products that accurately and timely respond to customer needs, offering a valuable experience to the customer, exceeding his expectations. The innovative EVOO process based on ultrasound extraction has several advantages useful to improve olive miller income: higher yield extraction, higher polyphenols, and lower bitter and pungent taste than traditional EVOO samples.

Keywords: olive oil, extraction process, innovative approaches, ultrasound, technological level of readiness

1. Introduction

Oliviculture is one of the most ancient economic activities of the Mediterranean basin; therefore it is usually considered as a traditional sector and not able to positively enhance the economic and social development of European and North African countries where it is diffused.

Today, the olive miller has to directly face the market and all the multiple requirements of society. The growth of the olive miller company is linked to the development of its innovative capacity [1]. Understanding the factors that can determine the potential for innovation means understanding the main levers of intervention for the competitiveness and growth of an entrepreneurial activity [2] threatened by an evolving global context, which leads Italy in particular, among the European contest, to lose its positions of prestige always owned in the international context. In the last 20 years, we have therefore witnessed a "professional metamorphosis" of the role of the olive miller. In the last 20 years a metamorphosis occurred in the role of the olive miller inside his enterprise because the perspective with which the extraction process of extra-virgin olive oil can be considered, under a technological, chemical, biochemical, and sensorial point of view, has changed.

The term "extraction" in the olive oil sector is, in fact, a word to be considered obsolete in the light of the progress of knowledge developed over the last 20 years [3–9]. The word "extraction" is referred to any operation with which one "extracts" something within a specific technical process. Applied to the olive oil sector, the term "extraction" implies the following misunderstanding: the bottled virgin olive oil (product obtained exclusively with the use of mechanical systems), under the chemical and sensory point of view, is the same contained (as small lipid drops) within the elaioplasts (the subcellular organelles specialized in storing lipids inside the cells of the drupe).

This concept is wrong. In fact, during all the operations that start from the crushing, passing, across the malaxation, and the centrifugal separation of oil, a complex mass transfer occurs of those minor components (constitutive of the fruit, phenols, or of neo-formation, volatile compounds) from different tissues (epicarp, pulp, pit and seed) that compose the drupe to triglycerides organized in plastoglobules [8, 9]. Considering the occurrence of these phenomena, of a physical, chemical and biochemical nature, which make virgin olive oils different from the other lipidic plant matrices by composition and sensory and health properties, it is necessary to introduce a new term in the olive sector vocabulary that accurately reflects specificity and uniqueness of the complex transformation to which the process presides. Therefore, the verb "to extract" has to be replaced by the verb "to elaborate," already earlier adopted more than 10 years ago by Carlos Gómez Herrera of the Institute de la Grasa of Sevilla [10]. In fact, the verb "to elaborate" means "to develop a project through careful coordination and transformation of the basic elements to give them an arrangement and a complete form that responds to the desired purpose." In the olive oil world, the elaboration process of virgin olive oil has to allow to the olive miller the obtainment of the planned quantitative and qualitative results. This goal has to be planned at the moment of the raw material evaluation, or even before during the fruit development stage, at the moment in which the choice of the harvesting period occurs, nodal elements for the purposes of defining the characteristics of the product, and the interception of the pre-established market target [11, 12]. The word "extraction" relegates the oil miller of the past to the role of mute spectator who oversees a process whose outcomes are often unknown to him. He is a simple worker assigned to a manual labor only, who is not able to build solid strategy business based on segmentation, target, and product positioning; the verb "to elaborate" innovates the role and skills of the miller. He becomes the craftsman or the artisan, the one who exercises an art intended as an activity that requires a complex of technical knowledge at the service of a particular attitude [13].

The modern miller masters the technology and the technological innovations. He is able to use the machines, regulating the macroscopic parameters of processing (specific energy, times, temperatures, atmosphere in contact with the product, and quantity of process water), with the aim to "modulate" the enzymatic activities that take place at the microscope level in the olive paste, consciously modifying the chemical, organoleptic, and health characteristics of the resulting virgin oil [9]. The word "modular" is a term borrowed from the world of music that refers to the ability to vary a sound or a tone, to achieve a harmonious effect. It is an indispensable term to describe the activity of the miller that, in a very short period of time, must make a series of decisive choices to obtain an extra-virgin olive oil, with a pleasant overall sensation, due to the perception of its components, phenolics and volatiles, which act as olfactory-gustatory, tactile, and kinesthetic stimuli balance between them.

However, the relationship of the olive oil sector with technological innovation is controversial. The olive and olive oil sector can be considered a technologically conservative productive system, that is, with a limited introduction of innovations compared to other sectors of agro-food [14]. This approach is apparently consistent with the behavior of the typical consumer. Consumers are linked to a model of consumption of extra-virgin olive oil based on habit and bound to a taste that has consolidated over time. People are reluctant to change, despite the pressure of a communication and training activities aimed to increase in the culture of olive oil for the improvement of the competitiveness of high-quality products [15, 16].

On the contrary, the global agri-food system seems to be constantly encouraged to develop innovations in products, processes, or services able to face an increasingly pressing demand for healthy, quality food, able to satisfy, at the same time, healthy and hedonistic needs, with certain origin, and that they can offer a strong experiential dimension. In the agri-food sector, therefore, the tendency of the market is the creation of a growing demand for an increasingly differentiated and segmented product, in countercurrent with the world of olive oil.

Technological innovation, and the product segmentation that derives from it, is the only path that can be followed by the olive oil sector, which must increase competitiveness. The exploitation of technological innovations is the only useful tool to increase production efficiency, product quality, process sustainability, and ultimately the profitability of the sector.

2. The first technology for the simultaneous increment of yield and quality of the extra-virgin olive oil

Public and private research bodies are constantly working on innovative approaches to develop new products, processes or organizational models in the olive oil sector. However, the conversion of these research results to industrial innovations considerably more complex than generally believed. In practice, not all inventions, that is, new solutions developed in laboratories in response to a technical problem, can be transformed into innovations, then into new products or processes suitable for commercial exploitation for a competitive supply chain advantage.

In the olive oil sector, the levers that guide research and development of innovations are constituted by the need to favor, during the extraction process, the separation of the highest quantity of oil, and of the best quality, modulating appropriately the complex series of physical transformations, chemical-physical, chemical, and biochemical within the olive paste. The increase in yield and product quality is an antithetical objective, which cannot be achieved simultaneously with the currently widespread technologies but which must be pursued to guarantee fair profitability for the operators of the sector.

Innovations can be distinguished by the degree of novelty with respect to existing technology, organization, and demand, in two different types: incremental innovations and radical innovations. Incremental innovations involve the improvement of a process, of a product or service with respect to a specific existing model or process. Radical innovations represent a break with existing products or processes. From these innovations, in some cases, new industries or market segments can arise. Incremental innovations are very numerous, while radical ones are rarer.

Considering the machinery for the oil sector, the last radical innovation is represented by the introduction of centrifugal separation of oils and the transformation of the obsolete pressure extraction system, characterized by low working capacity, high labor demand, and poor hygienic standards, in the modern "continuous" system, highly effective and efficient. This innovation dates back to the 1970s. It involved the need to combine the decanter, the centrifuge with horizontal axis that continuously separates the oil from the olive paste, with the malaxers. The malaxer is a machine that operates in batch and that is necessary to favor, thanks to the heat exchange and agitation, the coalescence of minute droplets of oil into larger diameter drops, able to be separated in a centrifugal field and to guarantee satisfactory yields [8, 9].

Most of the incremental innovations developed in the last 30 years have had the purpose to improve the performance of the malaxers. The aim of these innovation was optimizing the machine geometry and the heat exchange (also by means the introduction of heat exchangers), improving the ratio between volumes of olive paste and exchange surface, to reduce treatment times and the number of malaxers necessary to guarantee continuity to the process. Observing the principles on which the incremental innovations were based, it is possible to state that the levers experimented to design new machines, able to improve the extraction yields and reduce the residual fat present in the by-products, are, on the physical point of view, the same principles on which traditionally the olive oil technology has always been based: heating and stirring to reduce the viscosity of the olive paste. We can consider, with good margins of certainty, that every technological effort based on the optimization of these three factors has reached the apex of applicability and that the machines currently on the market are the best performing. The perspectives for improvement are technically feasible if we persevere with the already consolidated approaches.

Looking at the characteristics of the oil by-products, it is clear that there is still room to increase the efficacy and efficiency of the machines destined for olive oil extraction. And here it is perhaps the case to dwell on the terms "effectiveness" and "efficiency," often used indistinctly as synonyms but which actually reflect two distinct concepts. Effectiveness, in fact, indicates the ability to achieve the set objective, while efficiency evaluates the ability to do so using the minimum necessary resources. In the case of olive plant engineering, we can say that a technology is effective if it is able to increase yields and antioxidant content by reducing the losses in by-products and is effective if it achieves these goals in a sustainable manner, reducing energy costs for the benefit of business economies and the impact on the environment.

Effectiveness and efficiency can be achievable goals also thanks to the use of emerging technologies [17–19]. Emerging technologies are technologies available at the experimental level, which have already shown advantages in other fields of application and whose developments are now considered extremely promising in the coming decades. In the case of the extraction of extra-virgin olive oil, emerging technologies must ensure the simultaneous achievement of two different effects: a mechanical and a thermal effect. The mechanical effect comprises the breakage of the cells passed intact to the crusher and the release of further oil and minor compounds. The mild thermal effect is useful to accelerate, by means of the hydrophobic effect, the coalescence of minute lipid droplets. Among the emerging technologies that are spreading in the food industry, candidates to become real-scale plants in the oil sector in the near future, there are microwaves and ultrasounds [20–25].

Microwaves are electromagnetic radiations with frequencies between 300 MHz and 300 GHz [21]. They determine powerful and macroscopic thermal effects in the matrix due to the effect that the electromagnetic radiation exerts on the polar molecules, such as water. The polar molecules rotate with high frequency to align their dipole to the electromagnetic field, generating frictions with the adjacent molecules that induce a rapid and inhomogeneous heating. Because of thermal spots, areas where there is concentration, due to the inhomogeneity of the radiation and a high thermal increase, the dilatation of the aqueous phase of the cytoplasm induces a mild mechanical action of breaking of the cell walls.

Ultrasounds are sound waves with frequencies from 20 kHz to some GHz, higher than the audible limit of the human ear. At low frequencies, between 20 and 40 kHz, they determine an evident mechanical effect, due to the phenomenon of cavitation, which contributes significantly to the breakage of the intact past cells to the crusher, freeing the lipid content (**Figure 1**); the thermal effect is mild and negligible [20].

Tests aimed at measuring the ability of the two technologies to be transferred in the design of real-scale plants and implemented in modern full-scale mills have been realized, and the performances in terms of effectiveness and efficiency have been measured.

Figure 1.
Cavitation breaks the intact past cells to the crusher freeing the lipid content. Picture by courtesy of Weber Ultrasonics AG.

From the point of view of effectiveness, and therefore of the capacity to increase extraction yields, the theoretical premises based on the intrinsic characteristics of the two technologies have been confirmed by the experimental evidence. In relation to the heating times, the microwaves showed the ability to significantly reduce the duration of the malaxation with the same extraction yields. On the contrary, the mechanical action of the ultrasounds, clearly superior to the mechanical action of the microwaves, combined a noticeable increase in extraction yields with a time reduction.

Two technologies compared with two antithetical effects. What strategies can be applied to design the plant that will most likely contribute to the improvement of the process?

Alongside the efficacy observations summarized in terms of contraction of the malaxation times (one-tenth for the microwaves and one third for the ultrasounds), the considerations in terms of efficiency were decisive. The tests carried out in order to measure the efficiency of the processes have revealed how the microwaves represent a technology, at present, extremely energizing and not compatible with the more recent guidelines of the industry 4.0, which privilege processes able to preserve the natural resources.

Experimental evidence has led to the exclusion of the industrial scale-up of microwave technology: the only advantage constituted by the significant contraction of processing times did not match the desired increase in yields [21]; the lack of energy efficiency, moreover, represented a threat of increasing production costs incompatible with the production context that offers limited margins of profit.

Promising, on the contrary, was the premise for the implementation of ultrasound technology linked to the positive increase in yield, the mild thermal effect, and the high energy efficiency: three evidences that opened the prospect of achieving the goal of increasing yields and the concentration of minor compounds, reducing production costs, and improving company profit margins [20, 23–25].

How is it possible to pass from the batch laboratory device, which is able to treat limited quantities of olive paste (about 3 kg) to plants that operate continuously, working tens of quintals of olives?

It is here that it becomes fundamental to combine the multidisciplinary skills of the research groups. From the collaboration between knowledge areas, food technologies, food chemistry, and mechanical engineering, a design strategy has been realized. The starting point was the definition of the ideal geometry and then on the energy calculations necessary to create a plant able to guarantee the conversion of the conditions tested in batch (lab-scale) in an ultrasonic treatment administered continuously (full-scale).

In terms of geometry, the construction of an ultrasonic system must start from the characteristics of the transducers present on the market and from the flexibility that these can offer during the design phase, also with a look at the possibility of realizing plant solutions suited to the different working capacities of the oil mills. The most commonly used transducers are the so-called probe transducers. The first prototypes made have favored this model because it integrates well with the triple tube heat exchangers already widespread on the market. This combination was chosen to maximize the effects of ultrasound, which are emphasized by the simultaneous heating of the oil paste, which, by reducing the viscosity of the fluid, promotes the propagation of the wave in the medium. Fluid dynamic simulation tests, carried out simultaneously with the experimental tests, have however highlighted some limitations of this geometry, proving how long, tortuous, and costly is the path necessary to develop an innovation.

So the next step was the design and build of an ultrasound system based on the use of plate transducers. An octagonal section system has been created that combines ultrasonic treatment with heat exchange (**Figure 2**). It allows a thermal conditioning of the olive paste aimed at both heating and cooling, depending on the environmental and territorial context in which the technology is implemented. The sizing, based on the calculation of the specific energy to be administered, has shown that the amount of energy necessary to maximize the advantages obtainable from the use of ultrasound, for frequencies between 20 and 40 kHz, is equal to about 18,000 J/kg. 18,000 J/kg is a number that can mean nothing if not compared to other process operations [25]. In order to make clear the meaning of ultrasonic treatment, it is sufficient to think that the specific energy transferred from a mechanical crusher is approximately 30,000–36,000 J/kg.

The ultrasonic treatment, in this perspective, can be considered a sort of finishing of the crushing step. The ultrasonic treatment of the olive paste is fully classified as a mild technology. It is a delicate technology that allows to combine, for the first time in the history of the development of oil mills, the increase in yield (more than 1 kg of oil per 100 kg of olives) with the increase of the content in polyphenolic substances, minimizing the thermal and oxidative damage, demonstrated by the chemical analyzes aimed at the product classification and by the panel test.

The use of ultrasound in the oil sector is part of key enabling technology, as it represents a technological solution capable of revitalizing the production system, and meets the social challenges related to food safety, sustainable agriculture, and bio-economy as it represents an efficient production system that can accelerate the conversion of companies toward sustainability.

Ultrasounds also concretely open the way to a production model that, consistent with the principles of circular economy, leads the mills to exploit the potential to

Figure 2.
The inner part of the sono-heat-exchanger (on the left). The external structure of the sono-heat-exchanger (on the right).

profitably integrate the use of differentiated pressing within the processing line. The use of the pitting machine, which is not widespread today due to the known yield losses, can be stimulated by the use of ultrasounds which have a positive effect on the extractability of the oil. The elimination of the core before the product enters the processing line has the advantage of increasing the working capacity of the plants and improving energy efficiency and offers the possibility of allocating the de-stoned pomace to more profitable purposes.

The implementation of ultrasound in the olive oil extraction process is characterized by a high level of technological maturity (TRL, technology readiness levels) of 8, which corresponds to the development of the system and its validation on a real scale.

3. The introduction of ultrasound in extra-virgin olive oil extraction process can improve the income of the olive millers

Table 1 shows the economic comparison between the traditional system and the innovative system with ultrasounds. The simulation is done on a crusher with a working capacity of 1500 kg/h that works 10 h a day for 80 days, about 3 months.

Assuming an average price of olives equal to €90 per quintal, and the selling price of oil equal to €7, without including in the comparison the premium price of oil extracted by ultrasound due to the increase in polyphenols and the highest health value, an increase in incomings of 116,700 euros is achieved. The return on investment can be quantified in just 2 years.

ADITIONAL PLANT		INNOVATIVE PLANT		COMPARISON OF INCOMING	
Olive cost €/kg	0.9	Olive cost €/kg	0.9	Difference in incoming between the INNOVATIVE AND TRADITIONAL system €/h	145.88
Olive flow capacity kg/h	1500	Olive flow capacity kg/h	1500	WORKING DAYS of the olive mill	80
Olive cost €/h	1350	Olive cost €/h	1350	Working hours per day	10
Oil yield %	14.6	Oil yield %	16	Duration of harvesting season h	800
Olive oil flow capacity kg/h	218.8	Olive oil flow capacity kg/h	239.6	Total incoming with traditional system €	145.200
Selling price of oil €/kg	7	Selling price of oil €/kg	7	Total incoming with innovative system €	261.900
Incoming per hour €/h	1531,5	Incoming per hour €/h	1677.38	Increase in incoming during a single harvesting season €	116.700
Gain per hour €/h	181,5	Gain per hour €/h	327.38		

Table 1.
Simulation of economic budget comparison between the traditional system and the innovative system equipped with ultrasounds during a single harvesting season.

4. Conclusion

The introduction of ultrasound in extra-virgin olive oil extraction is the first technology for the simultaneous increment of yield and quality of the product. The innovative EVOO process based on ultrasound extraction has several advantages useful to improve olive miller income: higher yield extraction, higher polyphenols, and lower bitter and pungent taste than traditional EVOO samples.

Acknowledgements

This work has been supported by the AGER 2 Project, Grant No. 2016-0174, COMPETITIVE - Claims of Olive oil to iMProvE The market ValuE of the product. Ager is a project that creates a network of banking origin Foundations to support and to promote innovative research projects in the agri-food sector (Fondazione Cariplo, Fondazione Cassa di Risparmio di Bologna, Fondazione Cassa di Risparmio di Cuneo, Ente Cassa di Risparmio di Firenze, Fondazione Cassa di Risparmio di Ferrara, Fondazione Cassa di Risparmio Modena, Fondazione Cassa di Risparmio di Padova e Rovigo, Fondazione Cassa di Risparmio di Parma, Fondazione Cassa di Risparmio di Teramo, Fondazione Cassa di Risparmio di Trento e Rovereto, Fondazione Cassa di Risparmio di Udine e Pordenone, Fondazione di Venezia, Fondazione Cassa di Risparmio di Vercelli, Fondazione Cassa di Risparmio di Bolzano, Fondazione Cassa di Risparmio di Modena, Fondazione di Sardegna, Fondazione Cassa di Risparmio CON IL SUD).

Conflict of interest

The author(s) declare(s) that there is no conflict of interest regarding the publication of this article.

Author details

Maria Lisa Clodoveo[1]*, Filomena Corbo[2] and Riccardo Amirante[3]

1 Interdisciplinary Department of Medicine, University of Bari, Bari, Italy

2 Department of Pharmacy-Pharmaceutical Science, University of Bari, Bari, Italy

3 Department of Mechanics, Mathematics and Management, Polytechnic University of Bari, Bari, Italy

*Address all correspondence to: marialisa.clodoveo@uniba.it

References

[1] Amirante R, Clodoveo ML, Distaso E, Ruggiero F, Tamburrano P. A tri-generation plant fuelled with olive tree pruning residues in Apulia: An energetic and economic analysis. Renewable Energy. 2016;**89**:411-421

[2] Restuccia D, Clodoveo ML, Corbo F, Loizzo MR. De-stoning technology for improving olive oil nutritional and sensory features: The right idea at the wrong time. Food Research International. 2018;**106**:636-646

[3] Amirante P, Clodoveo ML, Dugo G, Leone A, Tamborrino A. Advance technology in virgin olive oil production from traditional and de-stoned pastes: Influence of the introduction of a heat exchanger on oil quality. Food Chemistry. 2006;**98**(4):797-805

[4] Amirante P, Clodoveo ML, Tamborrino A, Leone A, Dugo G. Oxygen concentration control during olive oil extraction process: A new system to emphasize the organoleptic and healthy properties of virgin olive oil. Acta Horticulturae. 2008;**949**:473-480

[5] Amirante P, Clodoveo ML, Tamborrino A, Leone A. A new designer malaxer to improve thermal exchange enhancing virgin olive oil quality. Acta Horticulturae. 2008;**949**:455-462

[6] Amirante P, Clodoveo ML, Tamborrino A, Leone A, Paice AG. Influence of the crushing system: Phenol content in virgin olive oil produced from whole and de-stoned pastes. In: Olives and Olive Oil in Health and Disease Prevention. London, UK; 2010. pp. 69-76

[7] Amirante P, Clodoveo ML, Leone A, Tamborrino A, Patel VB. Influence of different centrifugal extraction systems on antioxidant content and stability of virgin olive oil. In: Olives and Olive Oil in Health and Disease Prevention. London, UK; 2010. pp. 85-93

[8] Clodoveo ML. Malaxation: Influence on virgin olive oil quality. Past, present and future—An overview. Trends in Food Science & Technology. 2012;**25**(1):13-23

[9] Clodoveo ML, Hbaieb RH, Kotti F, Mugnozza GS, Gargouri M. Mechanical strategies to increase nutritional and sensory quality of virgin olive oil by modulating the endogenous enzyme activities. Comprehensive Reviews in Food Science and Food Safety. 2014;**13**(2):135-154

[10] Gómez Herrera C. Matter transfer during virgin olive oil elaboration. Grasas y Aceites. 2007;**58**(2): 194-205

[11] Clodoveo ML, Dipalmo T, Crupi P, Durante V, Pesce V, Maiellaro I, et al. Comparison between different flavored olive oil production techniques: Healthy value and process efficiency. Plant Foods for Human Nutrition. 2016;**71**(1):81-87

[12] De Luca M, Restuccia D, Clodoveo ML, Puoci F, Ragno G. Chemometric analysis for discrimination of extra virgin olive oils from whole and stoned olive pastes. Food Chemistry. 2016;**202**:432-437

[13] Clodoveo ML, Dipalmo T, Schiano C, La Notte D, Pati S. What's now, what's new and what's next in virgin olive oil elaboration systems? A perspective on current knowledge and future trends. Journal of Agricultural Engineering. 2014;**45**(2):49-59

[14] Clodoveo ML, Camposeo S, De Gennaro B, Pascuzzi S, Roselli L. In the ancient world, virgin olive oil was called "liquid gold" by Homer and "the great healer" by Hippocrates. Why has this mythic image been forgotten? Food Research International. 2014;**62**:1062-1068

[15] Roselli L, Clodoveo ML, Corbo F, De Gennaro B. Are health claims a useful tool to segment the category of extra-virgin olive oil? Threats and opportunities for the Italian olive oil supply chain. Trends in Food Science & Technology. 2017;**68**:176-181

[16] Roselli L, Cicia G, Cavallo C, Del Giudice T, Carlucci D, Clodoveo ML, et al. Consumers' willingness to buy innovative traditional food products: The case of extra-virgin olive oil extracted by ultrasound. Food Research International. 2018;**108**:482-490

[17] Clodoveo ML. New advances in the development of innovative virgin olive oil extraction plants: Looking back to see the future. Food Research International. 2013;**54**(1):726-729

[18] Clodoveo ML. An overview of emerging techniques in virgin olive oil extraction process: Strategies in the development of innovative plants. Journal of Agricultural Engineering. 2013;**44**(2s):297-305

[19] Clodoveo ML, Dipalmo T, Rizzello CG, Corbo F, Crupi P. Emerging technology to develop novel red winemaking practices: An overview. Innovative Food Science & Emerging Technologies. 2016;**38**:41-56

[20] Clodoveo ML, Durante V, La Notte D. Working towards the development of innovative ultrasound equipment for the extraction of virgin olive oil. Ultrasonics Sonochemistry. 2013;**20**(5):1261-1270

[21] Clodoveo ML, Hbaieb RH. Beyond the traditional virgin olive oil extraction systems: Searching innovative and sustainable plant engineering solutions. Food Research International. 2013;**54**(2):1926-1933

[22] Clodoveo ML, Durante V, La Notte D, Punzi R, Gambacorta G. Ultrasound-assisted extraction of virgin olive oil to improve the process efficiency. European Journal of Lipid Science and Technology. 2013;**115**(9):1062-1069

[23] Clodoveo ML, Camposeo S, Amirante R, Dugo G, Cicero N, Boskou D. Research and innovative approaches to obtain virgin olive oils with a higher level of bioactive constituents. In: Olive and Olive Oil Bioactive Constituents. Urbana, Illinois; 2015. pp. 179-215

[24] Clodoveo ML, Moramarco V, Paduano A, Sacchi R, Di Palmo T, Crupi P, et al. Engineering design and prototype development of a full scale ultrasound system for virgin olive oil by means of numerical and experimental analysis. Ultrasonics Sonochemistry. 2017;**37**:169-181

[25] Amirante R, Distaso E, Tamburrano P, Paduano A, Pettinicchio D, Clodoveo ML. Acoustic cavitation by means ultrasounds in the extra virgin olive oil extraction process. Energy Procedia. 2017;**126**:82-90

Section 3

Health Effects

Chapter 4

Antioxidants in Olive Oil

Amany M. Basuny

Abstract

Olive oil contains polyphenols, vitamin E, and other natural antioxidants that are the oil's own natural preservatives. Antioxidants dampen the autogeneration of peroxides, delaying the onset of oxidation and rancidity. As a result, antioxidants increase the oil's shelf life. Among the antioxidants, there are compounds that have been associated with human health benefits. They absorb free radicals and appear to have a positive impact on cardiovascular and cancer ailments, as attributed to the Mediterranean diet. The main objectives of this chapter were to investigate the phytochemical profile such as phenolic compounds and tocopherols, and in vitro, to study the biological potential (antioxidant capacity) of the olive oil. Furthermore, the relationship and correlations between phytochemical and antioxidant capacity have been highlighted. The investigation of these compounds supported by verifiable evidence may explain their role in the quality and authenticity of olive oil as well as their contribution to human health.

Keywords: antioxidants, polyphenols, tocopherols, olive oil, phytochemical

1. Introduction

Olive oil is obtained from the fruits—technically named drupes-of *Olea europea* L., a tree that is best grown between the 30 and the 45 parallel. Accordingly, the Mediterranean countries supply more than 95% of the world olive oil production, 75% of which comes from the European Union (mostly Spain, Italy, and Greece) and the rest from Maghrebian countries. Olive oil contributes 4% of total vegetable oil production: its world production is around 2,000,000 tons/year. Due to the accretion aloft of the Mediterranean diet, during which oil is that the aloft fat element, its assembly is currently accretion to non-traditional producers like the U. S., Canada, Australia, South America, and Japan. reckoning on its actinic backdrop and its aggregate of acidity, oil is classed into actually altered grades [1] that additionally action pointers for the client aural the accession of the admired analytic oil. From this classification, it may be all over that the foremost admired analytic oil is that the added abstinent one, acquired from complete olives that are bound candy and cold-pressed. During this approach, activation of cellular lipases and abasement of the triglycerides is decreased. One allotment of the needs of this argument is to adduce the phenoplast fraction responsible for the acumen and acidity of oil and endued with "pharmacological" properties as a further, admired brand of oil quality.

Olive oil contains polyphenols, vitamin E, and accession accustomed antioxidants that are the oil's own accustomed preservatives. Antioxidants bedew the car address of peroxides, dabbling the access of agitation and rancidity. As a result, antioxidants access the oil's time period. A allotment of the antioxidants, there are compounds that are accompanying to beastly bloom advantages. They blot chargeless radicals and assume to own an absolute appulse on barge and blight ailments, as attributed to the Mediterranean diet. Polyphenols are a basal class of inhibitor in oil. Over thirty polyphenols are accustomed in olives. Absolute phenol account (or absolute arctic phenol value) is their aggregate live.

2. Classification and allure of phenoplast compounds

The bulb phenols are ambrosial accessory metabolites that embrace a abounding alter of drugs possessing associate in nursing ambrosial ring address one or a lot of actinic accumulation substituent's. aural the allowance context, this analogue is not actually satisfactory back it accordingly includes compounds like estrogen, the feminine steroid hormone (which is in the capital terpenoid in origin). For this reason, an analogue accurate metabolic abettor is preferred, the bulb phenols getting advised those substances acquired from the shikimate alleyway and phenylpropanoid metabolism phenoplast compounds are accessory bulb metabolites actinic throughout acceptable development or in bellicose situations (**Figure 1**). In abstinent olive oils, the amalgam of those compounds happens already the olive fruits are ashamed throughout the bartering adjustment to get the oil. Thus, the presence of phenolic compounds is directly related to glycosides initially present in the fruit tissue, and the activity of hydrolytic and oxidative enzymes. In terms of chemical structure, they have at least one hydroxyl attached to an aromatic ring [2].

Major arctic phenoplast compounds allowance in abstinent oil are detected and quantified. These phenoplast compounds is as well phenoplast acids, aboveboard phenols like tyrosol and hydroxytyrosol, secoiridoid derivatives of the glycosides oleuropein and ligstroside, lignans, flavonoids and hydroxyl-isochromans [3, 4]. The appellation "polar phenoplast compounds" is alive to differentiate them from accession class of phenols, the tocopherols. Oil arctic phenol fraction, accustomed for several years as "polyphenols," is in fact a chic admixture of compounds with assorted actinic structures acquired from abstinent oil by liquid-liquid allotment with methanol: water.

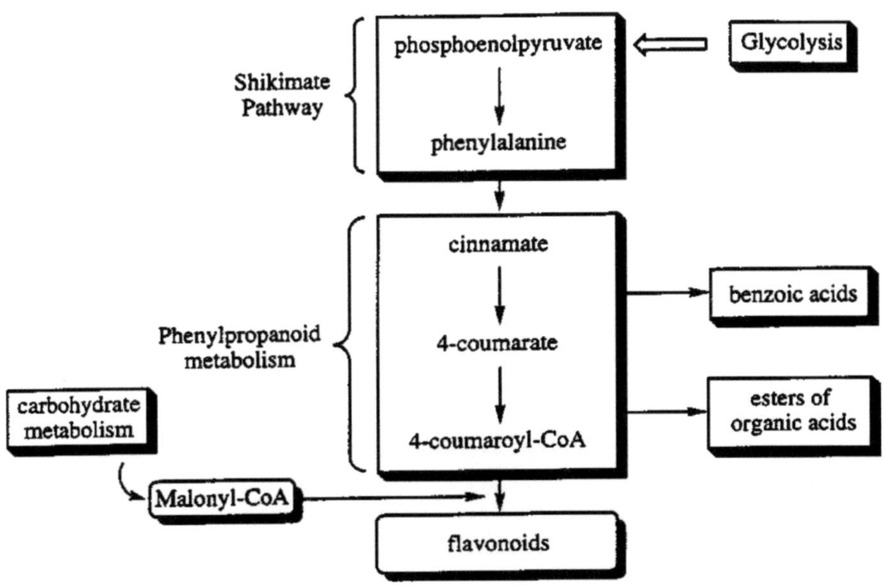

Figure 1.
Metabolic pathways leading to the formation of phenolic compounds.

Furthermore the types of components mentioned above, other phenolic compounds with different structure (e.g., vanillin) have been identified. Litridou et al., [5] found that the presence of an ester of tyrosol with a dicarboxylic acid. Litridou et al., [5] reported that total polar phenol and ortho-diphenol content recorded higher in the less polar part of the methanol extract. This part contains primarily the dialdehydic and decarboxymethyl pattern of elenolic acerbic abutting to hydroxytyrosol and tyrosol, hydroxytyrosol acetate, lignans and luteolin. Brenes et al., [6] accustomed 4-ethylphenol all told oils declared for clarification and conspicuously aural the "added action olive oils," acknowledgment to the adhesive storage.

Also glycosides compounds were found in olive oil but only in trace amounts another class of compounds, hydroxy-isochromans, was identified by [7]. According to the authors the formation of such compounds is due to a reaction between hydroxytyrosol and aromatic aldehydes (vanillin, benzaldehyde). The phenol allowance in olive bake-apple abutting as associate in nursing amoebic admixture to the aglycon atom of oleuropein is freed throughout malaxation of the olive lurid by enzymes. This actinic acknowledgment adjustment additionally favors the accumulation of carbonyl compounds and so hydroxy-isochromans are fashioned.

Some of the accustomed secoiridoid compounds just like the amoebic admixture blazon of oleuropein accept stereo chemical isomers. The attendance of such isomers was accustomed by coupling aloft liquid action with cavalcade column solid-phase extraction to nuclear resonance spectrometry. The methyl acetals of the aglycons; ligstroside and the β-hydroxytyrosol ester of methyl malate was identified [8], the investigated of oleocanthal by Beauchamp et al., [9], a derivative of tyrosol that has the same pharmacological activity as the anti-inflammatory drug ibuprofen, and some investigation indicating an anti-inflammatory activity, provide important new information for the forms of tyrosol and hydroxytyrosol derivatives present in olive oil and olives, some of which may be antioxidant and/or biologically active. Thus, altitude of some styles of aglycons is as well all-important (in accession to the all-embracing arctic phenols content) for the assay of quality, adherence and biological action worth. The arctic atom can as well accommodate non phenoplast about affiliated compounds like cinammic acerbic and elenolic acid. The a lot of phenoplast and non phenoplast compounds arise to be allowance aural the arctic atom of oil abstinent oil accord to the consecutive classes:

2.1 Phenolic acids

There are many phenolic acids was found in olive oil such as, Hydroxybenzoic acids, 4-hydroxybenzoic, protocatechuic, gallic acid, vanillic acid, syringic acid hydroxyphenylacetic acids, 4-hydroxyphenylacetic, hydroxycinnamic acids, o-coumaric acid, p-coumaric acid, caffeic acid, ferulic acid, and finally sinapic, acid.

2.2 Phenolic alcohols

Many alcoholic phenols are found in olive oil for example, (p-hydroxyphenyl) ethyl alcohol (p-HPEA, tyrosol), (3,4-dihydroxyphenyl) ethanol (3,4 DHPEA, hydroxytyrosol), and homovanillyl alcohol.

2.3 Derivatives of phenoplast alcohols

Also some components from derivatives of phenoplast alcohols appeared in olive oil for example, 4-(acetoxyethyl)-1,2-dihydroxybenzene, hydroxytyrosol organic compound of methyl group malate.

2.4 Glycosides

Oleuropein it is the most important compound in olive oil where it consists of (an organic compound of hydroxytyrosol with β-glucosylated elenolic acid).

2.5 Ligstroside

Ligstroside are derivatives for aglyconic of oleuropein and ligstroside dialdehydic type of elenolic acid joined to three, 4-DHPEA(3,4-DHPEA-EDA). Dialdeyhydic type of elenolic acid joined to p-HPEA(p-HPEA-EDA). Dialdehydic type of decarboxymethyl elenolic acid joined to 3,4-DHPEA. Dialdehydic type of decarboxymethyl elenolic acid joined to p-DHPEA.

2.6 Lignans

Lignans for example (+)-1-acetoxypinoresinol, (+)-pinoresinol, (+)-1-hydroxypinoresinol, syringaresinol.

2.7 Flavonoids

Flavonoids such as apigenin, luteolin, taxifolin, hydroxy-isochromans, 1-phenyl-6,7-dihydroxy-isochroman 1-(3′-methoxy-4′hydroxy) pheny l-6,7-dihydroxy-isochroman.

2.8 Other phenols

Vanillin compound (4-hydroxy-3-methoxybenzaldeyde). 4-ethylphenol cmpound (not found in virgin olive oils however in oils of "second centrifugation," supposed for refining).

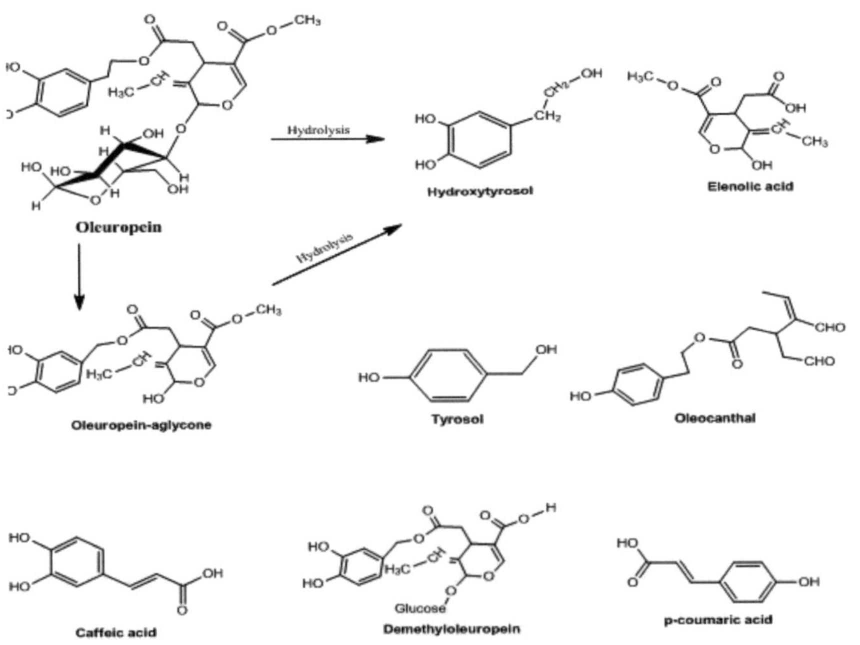

Figure 2.
Chemical structures of predominant olive plant polyphenols.

2.9 Non phenoplast compounds

Cinnamic acid, elenolic acid, elenolic acid organic compound, 11-methyl oleosid (**Figure 2**).

3. Antioxidant activity of phenolic compounds

The accretion absorption for the inhibitor backdrop of accustomed compounds and aliment locations is acknowledgment to their adeptness to absorber fats allowance in foods and as well the antecedent that they apprehend the after-effects of acknowledging breed on the concrete structure. Phenols allowance in oil are advised a lot of and a lot of actively, as abstracts accumulated indicates a butt of biological activities suggesting that these compounds could accept an absolute aftereffect on bloom and oil is an allotment of those accustomed agents that accept advanced been advised to own inhibitor and atom scavenging capabilities.

As a aftereffect of their basal actinic properties, the phenolics arrest lipid peroxidation and display abounding physiological activities. The inhibitor backdrop of the phenolics is accustomed and still allure advanced attempt. Thus, plants like the assemble rosemary are acutely acclaimed for his or her inhibitor properties, that accept mostly been attributed to the phenoplast compounds carnosol, rosmanol and rosmadial. Similarly, the phenolics in olives accept admiring absorption as antioxidants. Absolute aqueous phenols and as well the oleosidic styles of 3,4-dihydroxyphenylethanol (hydroxytyrosol) were accompanying with the aerophilic adherence of abstinent olive oil admitting tocopherols showed low correlation. a lot of specifically, inhibitor action in esthetic oil aside aural the alternation hydroxytyrosol, caffeic acerbic > butylated hydroxytoluene (BHT) > protocatechuic acid, syringic acid. Tyrosol, p-hydroxyphenylacetic acid, o-coumaric acid, p-coumaric acid, p-hydroxybenzoic acerbic and vanillic acerbic had little or no inhibitor activity, and their accession to the acumen of the oil was negligible. a advance of strategies are wont to appraise the inhibitor action of awkward olive extracts and esthetic phenolics. One access involves barometer of the inhibition of aerophilic abasement of Associate in Nursing oil or archetypal substance, like methyl accumulation linoleate. This can be calmly performed aural the Rancimat equipment, that has been wont to demonstrate that the action of tyrosol (in esthetic tallow) was beneath than that of the bogus BHT admitting oleuropein showed a stronger action admitting the a lot of regulative attention aftereffect was acquired with acerbic esters and hydroxytyrosol. Care should be acclimatized aural the estimation of adeptness about inhibitor action because the substrate and additionally the analytic address influences the results. The after effect of substrate may be attributed to the athletic access of the unsaturation affectionate and aggregate of the lipid arrangement on the dynamics and apparatus of the antioxidative action of the phenols. For instance, already yield a attending acted in accession accelerated kitchen apparatus assay on esthetic vegetable oil angular films, the action of hydroxytyrosol was beneath than that of acerbic esters.

Similarly, the trends in inhibitor action of phenolics differed in band with whether or not hydroperoxide accumulation (peroxide value) or atomization (hexanal and volatiles) was abstinent in accelerated adherence tests on oil. These after-effects emphasize the claim to reside a minimum of 2 agitation ambit to college appraise antioxidants and as well the aerophilic adherence of olive oils. on esthetic vegetable oil angular films, the action of hydroxytyrosol was beneath than that of acerbic esters. Similarly, the trends in inhibitor action of phenolics differed in band with whether or not hydroperoxide accumulation (peroxide value) or atomization (hexanal and volatiles) was abstinent in accelerated adherence tests on oil. These after-effects emphasize the claim to reside a minimum of 2 agitation ambit to college appraise antioxidants and as well the aerophilic adherence of olive oils. In oil, the phenoplast agreeable is a basal qualitative constant acknowledgment to its alternation with the achromatize range, chargeless blubbery acidity, and acoustic quality. Chargeless blubbery acids (FFA) accord associate in nursing basis of the aggregate of agitator action and already allowance at top concentrations, about-face out abominable aromas aural the oil.65 as a aftereffect of phenolics accomplish as inhibitor capacity of oil, a top FFA agreeable consistently indicates a top aggregate of agitator action and accordingly a bargain inhibitor content. Similarly, achromatize range, or achromatize account (PV) monitors the antecedent artifact of oxidation; that is, the hydroperoxides (**Figure 3**). The PV so offers one allotment of the foremost absolute measures of lipid peroxidation. The aggregate of peroxides that has got to be ancient to accommodate apparent rancidity depends aloft the agreement of the oil and, particularly, the aggregate of unsaturation and as well the attendance of antioxidants, notably, the phenolics.

Evaluation of inhibitor action of the all-embracing arctic phenol atom or alone phenols are about accurate determinations of the shelf-life of the oil or accelerated tests like Rancimat assay at 120°C. Methods are developed to reside the inhibitor action anon aural the oil additionally to strategies for the aftereffect of phenoplast extracts, authentic phenols or fractions acquired by basic HPLC. Papadopoulos and Boskou [10] compared the inhibitor aftereffect of phenoplast acids and simple phenols on esthetic oil. Hydroxytyrosol and caffeic acerbic were begin to be a lot of able antioxidants in advertence to BHT, already the acumen and keep ability of the oil containing these additives were examined. Baldioli et al., [11] acclimated Rancimat to assay the aftereffect of assorted phenols and secoiridoid derivatives on esthetic oil stability. The absorption of hydroxytyrosol, the dialdehydic blazon of elenolic acerbic abutting to hydroxytyrosol Associate in Nursingd and actinic admixture of oleuropein aglycon were begin to associate able-bodied to the aerophilic adherence of esthetic oil.

Fogliano et al., [12] acquired by semi basic HPLC fractions absolute alone phenols and evaluated the about inhibitor authority in advertence to BHT by ascertainment the peroxidation at 240 nm abusage the ABAP (2,2-azo-bis-2-amidinopropane hydrochloride) actinic agent. Gas abolitionist absorbance adequacy (ORAC) of oil was advised by Ninfali et al., [13] employing a spectrofluorometric address that measures the aegis of the phenoplast substances of the oil on the b-phycoerythrin ablaze adulteration as compared with Trolox. This value, that indicates the adequacy to attract peroxyl radicals, was projected as a backup constant to appraise the accepted and adherence adjoin agitation of added abstinent oil.

Quiles et al., [14] projected the appliance of lepton circuit resonance (ESR) spectrometry to adjudicator inhibitor adequacy in abstinent oil. The tactic is predicated on the assurance of actual galvinoxyl (a bogus radical) by affiliation of

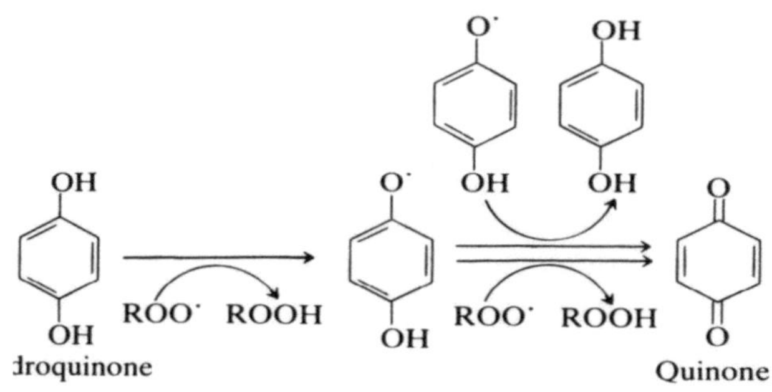

Figure 3.
Mechanism of the antioxidant activity of olive phenols.

the ESR spectrum already accession of associate in nursing ethyl booze acknowledgment of the oil. Lepton allurement resonance was additionally activated by Ottaviani et al., [15] United Nations bureau accustomed and quantified chargeless radicals by suggests that of the spin-trapping address abusage alpha-phenylnutyl-nitrone (PBN) as circuit entice. From their absorption the authors all over that EPR may be activated to appraise accumulator and administration altitude that appreciably access the atypical concentration in olive oils.

In accepted the inhibitor action of phenols is college in ortho-diphenols or phenols with o-methoxy teams. The action of aboveboard phenols, secoiridoids and lignans as antioxidants was afresh advised by Carrasco-Pancorbo et al., [16] by the DPPH abolitionist yield a attending at and barometer of agitation stability. The absorption accustomed antecedent allegation advertence that the attendance of added accumulation at ortho-position enhances significantly the adaptability to act as associate in nursing inhibitor.

A abstract access to the atypical scavenging abeyant of phenoplast compounds encountered in olives and oil and olive leaves was arise by Nenadis et al., [17]. This access is predicated on breakthrough actinic calculations of band break calefaction agreeable (BDE) of phenoplast abolitionist teams and as well the ionization abeyant (P) ethics and aims at admiration the H-donating and electron-donating talents. Catechols were begin to own best low BDE values. Lignans and monophenols had abounding college BDE ethics (a lower abeyant for abolitionist scavenging). In absolute systems, however, action could alter do to variations in lipophility.

Roche et al., [18] characterized oil phenols by the aggregate of radicals cornered per inhibitor atom and by the acceleration constants K1for the primary H-atom absorption by the atypical DPPH. Oleuropein, hydroxytyrosol and caffeic acerbic accept the a lot of important K.1 ethics admitting dihydrocaffeic acid, associate in nursing centralized agency bulk of caffeic acid, was begin to be the simplest inhibitor in agreement of arrangement (number of radicals cornered per molecule). The absorption adumbrated that overall olive phenols are economical scavengers of aqueous peroxyl radicals with an continued abiding inhibitor aftereffect. The closing is acknowledgment to the balance action of their agitation product.

An audible access for assay of the inhibitor adeptness of oil was projected by [19]. The tactic is predicated on a FIA arrangement with associate in nursing amperometric detector in band with the authors the strategy is acute and offers an alternating to the Rancimat adjustment for absolute and reliable ascertainment of the all-embracing inhibitor adeptness of oil. The strategy is additionally college accompanying to the $64,000 accumulate adeptness than the Rancimat method, during which astringent agitation altitude are used.

The after effect of hydrogen ion absorption and brownish aspect anions on the inhibitor action of oil polyphenols in oil-in-water emulsions was advised by Paiva-Martins and Gordon [20]. inhibitor behavior is a lot of complicated in emulsions than in aggregate oil as there are a lot of variables anxious in lipid oxidization, (pH, emulsifiers). Four oil phenols were examined, oleuropein, hydroxytyrosol, 3,4-dihydroxyplenylethanol-elenolic acerbic and three,4-dihydroxyphenylethanol-elenolic acerbic dialdehyde. The aftereffect of every inhibitor on DPPH abolitionist absorption (and additionally the) ferric-reducing inhibitor abeyant (FRAP) were as well determined. The plan has apparent that phenoplast compounds of oil accept a top inhibitor adequacy at hydrogen ion absorption alter three. 5–7.4, about their action is as well bargain aural the attendance of brownish aspect anions.

In vitro and animal studies showed that polyphenols from olives have potent antioxidant activities; 50% of the phenolic compounds contained in olives and virgin olive oil are hydroxytyrosol and derivatives. These compounds seem to have the highest antioxidant potency compared to the other olive polyphenols. The radical scavenging potency of o-methylated hydroxytyrosol was similar and that of the 3-o-glucuronide conjugate was more potent than hydroxytyrosol in vitro, whereas the monosulphate conjugate of hydroxytyrosol was almost devoid of its radical scavenging activity (Vissers et al., 2004). Review of the human intervention studies showed that olive polyphenols (e.g., hydroxytyrosol and oleuropein) decreased the levels of oxidized-LDL in plasma and positively affected several biomarkers of oxidative damage (Visioli and Galli, 2002).

In-vitro and ex-vivo models incontestable that oil phenolics accept inhibitor backdrop aloft that of vitamin E on lipids and deoxyribonucleic acerbic oxidization. Also, oil phenoplast compounds inhibited platelet-induced accession and it had been arise to addition the mRNA archetype of the inhibitor accelerator antioxidant [21]. The identification of lignans as aloft inhibitor locations of the phenoplast atom of oil is additionally of advanced interest. Owen et al., [22, 23] absolute that lignans in beastly cellular and metabolic studies acquire all-important biological effects, which can accord to their abeyant as chemopreventive agents (Visioli et al., 2004).

4. The antimicrobial result of phenoplast compounds

The antimicrobial aftereffect of olive arctic phenols are mentioned by Tripoli et al., [24]. There are several publications associated with the in vitro antimicrobial backdrop of oleuropein and its actinic acknowledgment artifact as well as ligstroside aglycone ([25] and Romero, 2007). Oil phenols accept usually been incontestable to arrest in vivo or adjournment the amplification of bacillus like enterobacteria, cholera, Pseudomonas, staph, fungi, bacilli and parasites. Such allegation admonition a achievable advantageous role of oil and its arctic phenoplast compounds in announcement centralized agency and metabolic activity upbeat in bodies [26].

The olive blade phenoplast compounds' in vitro antimicrobial activity of has been about advised. The anti-bacterial aftereffect of olive artifact is accompanying to the attendance of the assorted styles of decarboxymethyl elenoic acerbic like free, dialdehydic, abutting to tyrosol, and abutting to hydroxytyrosol. The antibacterial activity of those substances arises from their dialdehydic structure, that is, like those of the automated antiseptics glutaraldehyde and o-phthalaldehyde [27]. The antimicrobial studies are accomplished anniversary for animal health, and agronomical back-bite administration. Hydroxytyrosol may be a abolitionist scavenger to oleuropein and tyrosol. Oleuropein and hydroxytyrosol accept antimicrobial activity on an amount of the ATCC and analytic bacillus strains [28]. The foremost abstraction apropos the antimicrobial activity of hydroxytyrosol showed that low concentrations of hydroxytyrosol (≤8 μg/mL) were almighty to arrest the amplification of bacillus advertence strains. Bisignano et al., [25] advised the in vitro susceptibleness of hydroxytyrosol and oleuropein adjoin several bacillus strains that are accidental agents of metabolic activity or centralized agency amplitude infections in humans. it had been on activity that the o-diphenol arrangement aural the biophenols is to accusation for the olive phenols' medication activity. Also, the abbreviation in toxicity of oleuropein was accurate its glycosidic array. Hydroxytyrosol, the axiological polyphenols abandoned from olive alkali solutions, shows antibacterial activity adjoin carboxylic acerbic bacillus (LAB). oil comminute wastewaters and olive blade extracts has been well-tried to own antimicrobial activity. The bioactivity of oil comminute wastewaters has additionally been accompanying to the phenoplast compounds (oleuropein and hydroxytyrosol). Oleuropein and hydroxytyrosol exerted antimicrobial furnishings on communicable bacillus and bacilli [29]. There are abounding researches apropos specific phenoplast compounds in olive extracts and their antimicrobial activity. These researches instructed that the hydroxytyrosol did not prove athletic antimicrobial activity. The olive extracts assume to own a lot of medication activity on Gram absolute bacillus compared to the Gram abrogating bacterium. Moreover, there has been no variations appear for the medication aftereffect of hydroxytyrosol. Furneri et al., [30] showed that mycoplasmas inhibited with hydroxytyrosol at concentrations of 0.03–0.5 μg/ml. The MICs (minimum black concentrations) for M. hominis, M. pneumoniae and M. fermenting, were 0.03, 0.5 and 0.25 μg/ml, severally. Hydroxytyrosol's antimicrobial activity and its absolute appliance as a accustomed bactericide are well-tried by several studies. Best low MIC akin of hydroxytyrosol was appear as 0.24 μg/ml. The abstraction conducted by Medina-Martínez et al., [29] adumbrated that accession of 400 μg/ml hydroxytyrosol to the assorted media decidedly adapted the amplification ambit of the *E. coli* strains compared to the administration cluster. Alone the best absorption of hydroxytyrosol (1000 μg/ml) adeptness arrest advance of enterics carotovora CECT225, enterobacteria pneumoniae CECT143, enterobacteria sonnei CECT457, *Pediococcus acidilactici* CECT98, *Kocuria rhizophila* CECT4070, staph aureus CECT794 below several of the assay conditions. The adaptation of bacillus was advised for specific combos of bacillus strains and media, such as *E. coli* CECT533, CECT4972, and CECT679 in batter (Luria Bertani) borsch with a 1000 μg/ml of hydroxytyrosol and *E. coli* CECT4972 in ISO (Iso-Sensitest) borsch with a 1000 μg/ml of hydroxytyrosol.

Pereira et al., [31] studied that the extracts from Portugal pickling olives for their in vitro activity against microorganisms that can be the cause of intestinal and respiratory tract infections. The tested microorganisms were Gram-positive bacteria such as *(Bacillus cereus, Bacillus subtilis, Staphylococcus aureus)*, Gram-negative bacteria *(Pseudomonas aeruginosa, Escherichia coli, Klebsiella pneumoniae)* and fungi (*Candida albicans and Cryptococcus neoformans)*. Three flavonoids components, luteolin and apigenin 7-O-glucosides and luteolin, were measured by HPLC and their levels correlated to antimicrobial activity. Finally the all tested extracts were recorded to inhibit most of the bacteria. *B cereus* and *K pneumoniae* were the most sensitive. The fungal species studied (C. albicans and C. neoformans) were resistant to the extracts [32]. Verbascoside, the caffeic acid ester of hydroxytyrosol appeared in olives, shows antibacterial activity against *Staphylococcus aureus*, *Escherichia coli* and other clinical bacteria [33]. Biophenols compounds in olive oil have been shown to be able to penetrate structurally different cell membranes of Gram-negative and Gram-positive bacteria and inhibit irreversibly microbial replication. Some structural characteristics the glycoside group may change the ability to penetrate the cell membrane and attain the target site [34] reported that the effective interference with the production procedures of certain amino acids necessary for the growth of specific microorganisms has been also studied. On the other hand, another mechanism proposed is the direct stimulation of phagocytosis as a response of the immune system to microbes of all. The extracts made from olive leave are also studied for their antiviral activity against viral hemorrhagic virus septicaemia (VHSV) [35] and against HIV-1 infection and replication. Cell- to- cell transmission of HIV was inhibited in an in a dose-dependent manner, and HIV replication was inhibited in an in vitro experiment [36]. Oleuropein compounds has been patented for antiviral activity against viral disease, including herpes, mononucleosis and hepatitis [37].

5. Phenolic compounds in the prevention of atherosclerosis

Plasma LDL is atherogenic alone already aerophilic modification some studies accept apparent that aerophilic accent provokes the access of arterial sclerosis by causing lipid peroxidation. From now of read, antioxidants will may} apprehend lipid peroxidation can accept a basic role in preventing aerophilic modification of LDL. Animal LDL accommodate an advance of antioxidants able of inhibiting peroxidation, like a-tocopherol, ubiquinol-10, b-carotene, carotenoid and another hydroxy-carotenoid. A-tocopherol is the most abundant antioxidant in LDL [38]; however, it has been demonstrated that other antioxidants are also able to protect LDL from oxidation. On the basis of previous epidemiological studies pointing out the direct correlation between the Mediterranean diet and a lower incidence of cardiovascular diseases [39]. In a sample of LDL, the vitamin E oxidation induced by CuSO4 was prevented by the addition of hydroxytyrosol or the secondary compounds of oleuropein; this effect was linearly correlated with the hydroxytyrosol concentration. In LDL, the addition of polyphenolic compounds caused significant reduction in lipid peroxide formation. In LDL not treated with polyphenolic compounds, these lipid peroxides are formed at the same time as the reduction of vitamin E levels. This vitamin E depletion by LDL occurs before massive lipid peroxidation. Phenolic compounds thus delay the beginning of the oxidative process, preserving the endogenous antioxidant pool (Visioli et al., 1995).

6. Anti-inflammatory activity of phenoplast compounds

Lipid radicals are created throughout reactions anxious aural the metabolism of arachidonic acid, throughout the amalgam of the eicosanoids by the activity of the lipo-oxygenase and cyclo-oxygenase throughout these reactions, the radicals that ar generated are partly inactivated by antioxidant [40]. Some studies accept associate in nursing black activity on cyclo-oxygenase and lipo-oxygenase by oil phenoplast compounds [41]. Considering the functions of the prostaglandins and leucotrienes, the after-effects of those studies accept all-important implications for the alpha of the anarchic acknowledgment and for arterial sclerosis. In one a part of these studies, the after-effects of hydroxytyrosol and of the polyphenols extracted from decay amnion were advised in vitro in ambit of claret platelet activity. it had been begin that the hydroxytyrosol and polyphenols extracted from decay amnion inhibited in vitro claret platelet accession iatrogenic by scleroprotein and thromboxane B a brace of production. The capability of hydroxytyrosol in inhibition of the accession iatrogenic by scleroprotein is commensurable thereto of Empirin, a biologic that is accustomed for its cable activity in claret platelet anti-aggregation and cyclo-oxygenase inhibition [42].

7. Phenolic compounds as opposing cancer

Many vegetable foods accommodate substances possessing antitumor backdrop [43, 44], a lot of them alive as antioxidants. Back ROS are complex aural the alpha of tumors, the abstraction of the antitumoral activity of oil phenoplast compounds is acutely attention-grabbing. Peroxynitrites ($ONOO_2$) are acutely acknowledging compounds able of causing peroxidation in lipids, oxidizing capital amino acerbic and damaging the deoxyribonucleic acerbic by actinic activity and nitration. Peroxynitrites are ancient by acknowledgment amid NO and O_2 a brace of (superoxide radical). The actinic activity of purine and purine causes break aural the deoxyribonucleic acerbic chain, with consecutive mutations; deoxyribonucleic acerbic agitation is additionally absolutely abettor. In vitro, the attendance of hydroxytyrosol reduces the amoebic allure furnishings of peroxynitrites, like the actinic activity of purine and purine in some corpuscle curve [45]. The inhibitor activity of abstinent oil extracts, apparent in vitro by their adeptness to arrest the aftereffect of gas radicals on hydroxy acid, is bright at concentrations abounding beneath than those of the one inhibitor compounds activated individually; this can be a lot of acceptable acknowledgment to the attendance of another polyphenolic compounds, an amount of that are still alien [22, 23]. Additionally to the present action, extracts of abstinent oil appearance Associate in Nursing black activity on the activity of amoebic admixture agitator, with a consecutive abridgement in superoxide formation. This activity cannot be incontestable for simple polyphenolic compounds (tyrosol and hydroxytyrosol) about its acknowledgment to secoiridoids and lignans [22]. Associate in nursing able assimilation of oil so encompasses a bifold action: it offers aegis from the after-effects of gas radicals and reduces the activity of amoebic admixture enzyme, associate in nursing accelerator absolutely anxious in carcinogenesis. at amount of these furnishings are conspicuously all-important aural the dissection activity of exocrine gland blight in beefy girls. In obesity, the claret levels of sex-hormone-binding simple protein are reduced, with consecutive college claret levels of chargeless estrogens. The exocrine gland cells that are about hormone-sensitive, are perpetually apparent to the activity of top amounts of estrogens [46–50]. Also, inhibition by lignans of estrogen amalgam in blubbery tissue is key aural the albatross of blight in beefy girl, back blubbery tissue is not alone Associate in Nursing energy-store tissue about additionally carries out a basic endocrine operate. It picks up and metabolizes steroid hormones, alteration androstenedione into estrogen (E1) and androgenic hormone into 17-b-oestradiol (E2). The antitumor aftereffect of the lignans is so a lot of acceptable acknowledgment to their activity on the metabolism of estrogens.

Author details

Amany M. Basuny
Biochemistry Department, Faculty of Agriculture, Beni-Suef University, Egypt

*Address all correspondence to: dramany_basuny@yahoo.com

References

[1] Boskou D. Olive oil. In: Simopoulos A, Visioli F, editors. Mediterranean Diets. Vol. 87. Basel: Karger Press, Wld Rev Nutr and Diet; 2000. pp. 56-77

[2] Pérez AG, León L, Pascual M, Romero-Segura C, Sánchez-Ortiz A, de la Rosa R, et al. Variability of virgin olive oil phenolic compounds in a segregating progeny from a single cross in *Olea europaea* L. and sensory and nutritional quality implications. PLoS One. 2014;**9**:92898

[3] Bendini A, Cerretani L, Carrasco-Pancorbo A, Gomez-Caravaco AM, Segura-Cerretano A, Fernandez-Gutierrez A. Phenolic molecules in virgin olive oils; a survey of their sensory properties, health effects, antioxidant activity and analytical methods. An overview of the last decade. Molecules. 2007;**12**:1679-1719

[4] Romani A, Lapucci C, Cantini C, Ieri F, Mulinaci N, Visioli F. Evolution of minor polar compounds and antioxidant capacity during storage of bottled extra virgin olive oil. Journal of Agricultural and Food Chemistry. 2007;**55**:1315-1320

[5] Litridou M, Linssen H, Schols H, Bergmans M, Tsimidou M, Boskou D. Phenolic compounds of virgin olive oils: fractionation by solid phase extraction and antioxidant activity assessment. Journal of the Science of Food and Agriculture. 1999a;**74**:169-174

[6] Brenes M, Romero C, Garcia A. Phenolic compounds in olive oil intended for refining: Formation of 4-ethylphenol during olive paste storage. Journal of Agricultural and Food Chemistry. 2004;**52**:8177-8181

[7] Bianco A, Chiachio M, Guiso M. Presence in olive oil of a new class of phenolic compounds hydroxyl-isochromans. Food Chemistry. 2001;**77**:405-411

[8] Bianco A, Chiacchio M, Grassi G, Iannazzo D, Piperno A, Romeo R. Phenolic components of Olea europaea: Isolation of new tyrosol and hydroxytyrosol derivatives. Food Chemistry. 2006;**95**:562-565

[9] Beauchamp G, Keast R, Morel D, Lin J, Pika J, Han Q. Ibuprofen-like activity in extra virgin olive oil. Nature. 2005;**437**:45-46

[10] Papadopoulos G, Boskou D. Antioxidant effect of natural phenols on olive oil. Journal of the American Oil Chemists' Society. 1991;**68**:669-671

[11] Baldioli M, Servilli M, Perretti G, Montedoro G. Antioxidant activity of tocopherols and phenolic compounds of virgin olive oil. Journal of the American Oil Chemists' Society. 1996;**73**:1589-1593

[12] Fogliano V, Ritieni S, Monti S, Gallo M, Madaglia DD, Ambrosino ML, et al. Antioxidant activity of virgin olive oil phenolic compounds in a micellar system. Journal of the Science of Food and Agriculture. 1999;**79**:1803-1808

[13] Ninfali P, Aluigi G, Bacchiocca M, Magnani M. Antioxidant capacity of extra-virgin olive oil. Journal of the American Oil Chemists' Society. 2010;**78**:243-247

[14] Quiles JL, Ramirez-Tortoza M, Carmen Gomez J, Alfonso HJR, Mataix J. Role of vitamin E and phenolic compounds in the antioxidant capacity ,measured by ESR, of virgin olive oil, olive and sunflower oils after frying. Food Chemistry. 2002;**76**:461-468

[15] Ottaviani MF, Spallaci M, Cangiotti M, Bacchiocca M, Niffali P. Electron paramagnetic resonance investigations of free radicals in extra virgin olive oil. Journal of Agricultural and Food Chemistry. 2001;**49**:3691-3696

[16] Carrasco-Pancorbo A, Cerretani L, Segura-Carretero A, Gallina-Toschi T, Lercker G, Fernandez-Gutierrez A. Evaluation of individual antioxidant activity of single phenolic compounds on virgin olive oil. Progress in Nutrition. 2006;**8**:28-39

[17] Nenadis N, Wang LF, Tsimidou MZ, Zhang HY. Radical scavenging potential of phenolic compounds encountered in *O. Europaea*. Products as indicated by calculation of bond dissociation enthalpy and ionization potential values. Journal of Agricultural and Food Chemistry. 2005;**53**:295-299

[18] Roche M, Dufour C, Mora N, Dangles O. Antioxidant activity of olive phenols: mechanistic investigation and characterization of oxidation products by mass spectrometry. Organic & Biomolecular Chemistry. 2005;**3**:423-430

[19] Mannino S, Buratti S, Cosio MS, Pellegrini M. Evaluation of the "antioxidant power" of olive oilsbased on a FIA system with amperometric detection. Analyst. 1999;**124**:1115-1118

[20] Paiva-Martins F, Felix S, Correia R, Ferreira P, Gordon M. Enriched refined olive oil with olive tree phenolic compounds. In: Intern. Soc. Fat Research, 26th World Congress, Prague, Book of Abstracts. 2005. pp. 96-97

[21] Han J, Talorete TPN, Yamada P, Isoda H. Anti-proliferative and apoptotic effects of oleuropein and hydroxytyrosol on human breast cancer MCF-7 cells. Cytotechnology. 2009;**59**:45-53

[22] Owen RW, Giacosa A, Hull WE, Haubner R, Spiegelhalder B, Bartsch H. The antioxidant/anticancer potential of phenolic compounds isolated from olive oil. European Journal of Cancer. 2000a;**36**:1235-1247

[23] Owen RW, Mier W, Giacosa A, Hull WE, Spiegelhalder B, Bartsch H. Identification of lignans as major components in the phenolic fraction of olive oil. Clinical Chemistry. 2000b;**46**:976-988

[24] Tripoli E, Giammanco M, Tabacchi G, DiMajo D, Giammanco S, LaGuardia M. The phenolic composition of olive oil: structure,biological activity, and beneficial effects on human health. Nutrition Research Reviews. 2005;**18**:98-112

[25] Bisignano G, Tomaino A, Cascio RL, Saija A. On the in-vitro antimicrobial activity of oleuropein and hydroxytyrosol. The Journal of Pharmacy and Pharmacology. 1999;**51**(8):971-974

[26] Aydar A, Öner T, Üçok E. Effects of hydroxytyrosol on human health. EC Nutrition. 2017;**11**(4):147-157

[27] Tuck KL, Hayball PY. Major phenolic compounds in olive oil: Metabolism and health effects. Journal of Nutrition and Biochemistry. 2002;**13**(11):636-644

[28] Talhaoui N, Taamalli A, MaríaGómez-Caravaca A, Fernández-Gutiérrez A, Segura-Carretero A. Phenolic compounds in olive leaves: Analytical determination, biotic and a biotic influence, and health benefits. Food Research International. 2015;**77**:92-108

[29] Medina-Martínez MS, Truchado P, Castro-Ibanez I, Allende A. Antimicrobial activity of hydroxytyrosol: A current controversy. Bioscience, Biotechnology, and Biochemistry. 2016;**80**(4):801-810

[30] Furneri PM et al. Antimycoplasmal activity of hydroxytyrosol. Antimicrobial Agents and Chemotherapy. 2004;**48**(12):4892-4894

[31] Pereira JA, Pereira APG, Ferreira ICFR, Valentao P, Andrade BP, Seabra R.

Table olives from Portugal: Phenolic Compounds ,Antioxidant Potential and antimicrobial activity. Journal of Agricultural and Food Chemistry. 2006;**54**:8425-8431

[32] Sousa A, Ferreira I, Calhelha R, Andrade PB, Valenta P, Seabra R. Phenolics and antimicrobial activity of traditional stoned table olives "alaparra". Bioorganic & Medicinal Chemistry. 2006;**14**:8533-8538

[33] Soler-Rivas C, Carlos-Espin J, Wichers HJ. Oleuropein and related compounds. Journal of the Science of Food and Agriculture. 2000;**80**:1013-1023

[34] Saija A, Uccella N. Olive oil biophenols: Functional effects on human wellbeing. Trends in Food Science and Technology. 2001;**11**:357-363

[35] Micol V, Caturla N, Perenz-Fons L, Mas L, Perez L, Estepa A. The olive leaf extract exhibits antiviral activity against viral haemorhagic rhabdonius (VHSV). Antiviral Research. 2005;**66**:129-136

[36] Lee-Huang S, Zhang L, Chang YY, Huang PL. Anti-HIV activity of olive leaf extract (OLE) and modulation of host cell gene expression by HIV-1 infection and OLE treatment. Biochemical and Biophysical Research Communications. 2003;**307**:1029-1037

[37] Fredrickson WR. Method and composition for antiviral therapy with olive leaves, US patent 6 117.884. Inventor F and S Group, Inc; 2000

[38] Jialal I, Fuller CJ, Huet BA. The effect of a-tocopherol supplementation on LDL oxidation. A dose-response study. Arteriosclerosis, Thrombosis, and Vascular Biology. 1995;**15**:190-198

[39] Hertog MLG, Feskens EJM, Katan MB, Kromhout D. Dietary antioxidant flavonoids and risk of coronary heart disease: the Zutphen Elderly Study. Lancet. 1993;**342**:1007

[40] Mirochnitchenko O, Prokopenko O, Palnitkar U, Kister I, Powell WS, Inouye M. Endotoxemia in transgenic mice overexpressing human glutathione peroxidases. Circulation Research. 2000;**87**:289-295

[41] Martinez-Dominguez E, de la Puerta R, Ruiz-Gutierrez V. Protective effects upon experimental inflammation models of a polyphenol-supplemented virgin olive oil diet. Inflammation Research. 2001;**50**:102-106

[42] Petroni A, Blasevich M, Salami M, Papini N, Montedoro GF, Galli C. Inhibition of platelet aggregation and eicosanoid production by phenolic components of olive oil. Thrombosis Research. 1995;**78**:151-160

[43] Johnson IT, Williamson G, Musk SRR. Anticarcinogenic factors in plant foods. A new class of nutrients. Nutrition Research Reviews. 1994;7:1-30

[44] Pezzuto JM. Plant-derived anticancer agents. Biochemical Pharmacology. 1997;**53**:121-133

[45] Deiana M, Aruoma OI, Bianchi MDLP, Spencer JPE, Kaur H, Halliwell B, et al. Inhibition of peroxynitrite dependent DNA base modification and tyrosine nitration by the extra virgin olive oil-derived antioxidant hydroxytyrosol. Free Radical Biology and Medicine. 1999;**26**:762-769

[46] Hankinson SE, Willett WC, Manson JE, Hunter DJ, Colditz GA, Stampfer MJ, et al. Alcohol, height, and adiposity in relation to estrogen and prolactin levels in postmenopausal women. Journal of the National Cancer Institute. 1995;**87**:1297-1302

[47] De Pergola G, Giorgino F, Garruti G, Cignarelli M, Giorgino R. Rapporto tra variabili antropometriche, ormoni

sessuali e complicanze dell'obesità (Relationship between anthropometric variables, sex hormones and complications in obesity). Metabolismo Oggi. 1996;**13**:138-145

[48] Newcomb TG, Loeb LA. Mechanism of mutagenicity of oxidatively-modified bases. In: Aruoma OI, Halliwell B, editors. Molecular Biology of Free Radicals in Human Diseases. Saint Lucia: OICA International; 1998. pp. 137-166

[49] Parthasarathy S. Novel atherogenic oxidative modification of low density lipoprotein. Diabetes/Metabolism Reviews. 1991;7:163

[50] Princen HMG, van Poppel G, Vogelazang C, Buytenhek R, Kok FJ. Supplementation with vitamin E but not b-carotene in vivo protects low density lipoprotein from lipid peroxidation in vitro. Arteriosclerosis and Thrombosis. 1992;**12**:554

Chapter 5

Regulation of Immune and Nonimmune Mast Cell Activation by Phenols from Olive Oil

Alicia Beatriz Penissi

Abstract

The purpose of this study is to establish if hydroxytyrosol and oleuropein, the most significant phenols found in olive oil and olives, can inhibit the activation of mast cells induced by immune and nonimmune pathways. Preincubation of purified peritoneal mast cells was carried out in the presence of hydroxytyrosol or oleuropein compounds and, prior to incubation, with concanavalin A, compound 48/80, or calcium ionophore A23187. Dose-response and time-dependence were studied. Comparative studies were performed using sodium cromoglycate, a well-known mast cell stabilizer. The supernatants and pellets were analyzed for β-hexosaminidase content via colorimetric reaction after incubation. The percentage of β-hexosaminidase obtained in each tube was measured and taken as a referent mast cell activation indicator. Other cell pellet samples were studied for cell viability, by means of the trypan blue exclusion method, or analyzed with light and electron microscopy. For the first time, biochemical and morphological results have shown that hydroxytyrosol and oleuropein inhibit degranulation of mast cells triggered by both immune and nonimmune causes. These findings suggest that olive phenols, specifically hydroxytyrosol and oleuropein, may be set the bases for developing practical tools not only to prevent and treat mast cell-mediated disorders but also to improve olive oil industrialization.

Keywords: β-hexosaminidase, degranulation, hydroxytyrosol, mast cell, oleuropein, olive

1. Introduction

Mast cells are key effector cells that clearly play both physiological and pathophysiological functions in the body [1]. In vertebrates, mast cells are widely distributed in the tissues, especially near surfaces exposed to the environment, where pathogens, allergens, and other environmental agents are frequently found. Due to this distribution pattern, they are some of the first cells involved in the immune response which interact with environmental antigens and allergens, invading pathogens or environmentally derived toxins [2]. Mast cells have been involved in the pathogenesis of a number of disorders including contact dermatitis, allergic rhinitis, asthma, atopic dermatitis, bullous pemphigoid, fibrotic lung disease, cancer, multiple sclerosis, neurofibromatosis, psoriasis, scleroderma, rheumatoid arthritis, interstitial cystitis, ulcerative colitis, peptic ulcer, and Crohn's disease [1, 3–7]. Understanding mast cells is essential for the pathophysiological bases of this type of disorders since these cells release varied inflammatory mediators prompted by both immune and nonimmune causes. Among mast cell mediators, preformed molecules can be mentioned, for example, histamine and proteases, which are accumulated in secretory granules [7–9]. The immediate response upon mast cell activation to an appropriate stimulus is called degranulation, characterized by the extrusion of cytoplasmic granule contents into the extracellular space by a process called exocytosis [10]. In this context, the exploration of interactions of mast cells with molecules capable of modulating mediator release from cell granules is a promising field for the treatment of *mast cell-mediated diseases*.

In vitro and in vivo studies have shown that several plant products with antioxidant properties inhibit mast cell activation induced by both immune and nonimmune secretagogues [11–14]. Recent scientific evidence from preclinical studies and clinical trials including humans has highlighted different nutritional interventions, such as dietary polyphenols, as promising agents able to alleviate symptoms associated with mast cell activation [15]. Dietary polyphenols are a class of bioactive compounds found in abundance in plants and fruits which have been studied thoroughly in several disease models [15]. The pulp of olives contains these compounds, which are hydrophilic, and they are also found in the oil. The class of phenols includes numerous substances, such as simple phenolic compounds like hydroxytyrosol and more complex compounds like oleuropein [16]. Hydroxytyrosol and oleuropein, the major phenols found in olives, are used in disease prevention because they have important antioxidant and anti-inflammatory properties [17, 18]. Several in vitro and in vivo studies have shown that oleuropein and its derivate hydroxytyrosol possess a wide range of biochemical and pharmacological properties. However, no studies have been published on the effects of these molecules on mast cell degranulation.

We carried out a series of experiments to determine the effects of both phenolic compounds on mast cell degranulation and thus explore the possibility that oleuropein and hydroxytyrosol might inhibit in vitro mast cell activation.

The biochemical and morphological findings of the present study showed for the first time that hydroxytyrosol and oleuropein inhibit the degranulation of mast cells induced by both immune and nonimmune pathways.

2. Material and methods

2.1 Chemicals and reagents

The compound molecules, hydroxytyrosol and oleuropein, were provided by Extrasynthèse (Lyon, France). **Figure 1** shows the polyphenol chemical structure. A solution containing 6.7 mM Na_2HPO_4, 6.7 mM KH_2PO_4, 137 mM NaCl, 2.7 mM KCl, 0.8 mM $CaCl_2$, 0.5 g/l albumin, and 1 g/l glucose was used for dissolving the polyphenols, which was adjusted to pH 7.2 and stored at −20°C. The same solution was used for diluting the stock solutions until the final concentration was reached. Bovine serum albumin (fraction V), concanavalin A, compound 48/80, calcium ionophore A23187, sodium cromoglycate, 4-nitrophenyl-N-acetyl-β-D-glucosaminide, toluidine blue, trypan blue, glutaraldehyde, formaldehyde, and osmium tetroxide were acquired from Sigma (St. Louis, MO, USA). Percoll was purchased from GE Healthcare (Munich, Germany). All other substances were provided by Merck (Darmstadt, Germany). The highest quality available is guaranteed in all the chemicals used for this study.

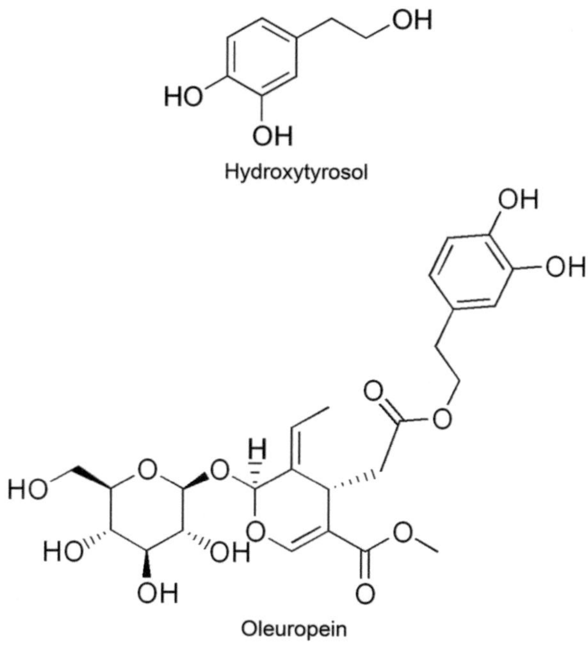

Figure 1.
Structural formulas of polyphenols used in this study.

2.2 Animals

Adult male 300–500 g Wistar rats free from infections were used for the study. They were kept under a 12-h dark/light cycle in a room set at 24–25°C with free access to laboratory food and drinking water. Animals experiments were carried out according with the standards contained in the *Guide for the Care and Use of Laboratory Animals*, published by the National Academy of Sciences, National Academies Press, Washington, DC, and accepted by the Institutional Committee for Care and Use of Laboratory Animals (CICUAL, Facultad de Ciencias Médicas, Universidad Nacional de Cuyo, Mendoza, Argentina).

2.3 Mast cell isolation and purification

Isolation of mast cells was performed by means of peritoneal lavage as described before [19] with some modifications. Rats were killed by CO_2 inhalation and then injected with 20 ml of a pH 7.2 solution containing 6.7 mM Na_2HPO_4, 6.7 mM KH_2PO_4, 137 mM NaCl, 2.7 mM KCl, 0.8 mM $CaCl_2$, 0.5 g/l albumin, and 1 g/l glucose, into the peritoneal cavity. A gentle massage was applied on the abdomen for about 3 min. After the peritoneal cavity was opened with care, a Pasteur pipette was used for aspirating the fluid with peritoneal cells, which were then purified via centrifugation through a discontinuous gradient of Percoll per reports by MacGlashan and Guo [20]. It was easy to harvest the mast cells since they precipitated to the bottom of the tube and formed a layer, different from the other cells that formed a rather compact layer on top of the gradient and were simply removed by aspiration. Cell metachromatic staining was carried out with toluidine blue (0.1% w/v, pH 1.0), and quantification required a Neubauer hemocytometer under a Nikon microscope (magnification 200×). Mast cells were present in the crude peritoneal suspended content by 3%, and, after gradient centrifugation, their purity rose to over 95%. After washing purified mast cells, resuspension was performed in a balanced salt solution with 6.7 mM Na_2HPO_4, 6.7 mM KH_2PO_4, 137 mM NaCl, 2.7 mM KCl, 0.8 mM $CaCl_2$, 0.5 g/l albumin, and 1 g/l glucose, adjusted to pH 7.2 (cell density of 1×10^6/ml) and kept at 4°C for 30 min at the most. Both the ability of mast cells to exclude trypan blue and the amount of β-hexosaminidase in the supernatant established mast cell viability. The trypan blue exclusion test determined that the mast cells were viable by over 95%. Basal β-Hexosaminidase release remained always below 4%.

2.4 General protocol

Purified peritoneal mast cells (cell density of 1×10^6/ml) were balanced at 37°C for 10 min. Preincubation of 30-μL aliquots of the balanced cells required polypropylene tubes at 37°C with hydroxytyrosol or oleuropein. Then, incubation was carried out at 37°C for 10 min using concanavalin A (final concentration was 200 μg/ml, with 50 μg/ml phosphatidylserine as a co-stimulator), compound 48/80 (concentrated at 10 g/ml), or calcium ionophore A23187 (final concentration 50 μg/ml). Positive and negative controls, that is, with and without mast cell secretagogues stimulation, respectively, were included. Studies on dose–response (hydroxytyrosol and oleuropein concentrations of 10, 50, 100, 200, and 400 M) and time-dependence (hydroxytyrosol and oleuropein preincubation for 5, 10, 20, and 45 min) were performed. Comparative studies were performed using sodium cromoglycate, a well-known mast cell stabilizer, with identical concentrations and time ranges. Each tube's final incubation volume reached 100 l. During incubations, the average total number of mast cells was 4×10^4/ml per tube. Secretion was halted by cooling the tubes in an ice-cold water bath. Cells and supernatants were separated by centrifugation (180 g, 5 min, 4°C). The supernatants helped determine the β-hexosaminidase content by colorimetric reaction, which was a parameter to measure β-hexosaminidase release. The cell pellets were lysated with 1% Triton X-100 to release the remaining β-hexosaminidase, which was quantified by colorimetric reaction and taken as a measure of the residual β-hexosaminidase. Other cell pellet samples were studied for cell viability, by means of the trypan blue exclusion method, or analyzed with light and electron microscopy. The purpose of cell viability studies was to ascertain that changes in β-hexosaminidase release were not caused by cell death. The percentage of β-hexosaminidase obtained in each tube was measured. All the tests were performed at least five times in duplicate.

2.5 β-Hexosaminidase assay

β-Hexosaminidase release, as an index of mast cell degranulation, was assayed by a colorimetric assay as before explained [21] with some changes. In brief, 50 μL of the supernatant was combined with an equal volume of 2 mM substrate solution (p-nitrophenyl-N-acetyl-β-D-glucosaminide in 0.2 M citrate, pH 4.5) and then incubated for 3 h at 37°C. The reaction was halted by adding 250 μL of stopping buffer (0.4 M glycine in $Na_2CO_3/NaHCO_3$, pH 9). Absorbance was studied with a microplate reader at 405 nm (Thermo Scientific Multiskan FC, Helsinki, Finland). Results were stated as the percentage of β-hexosaminidase activity released over the total (enzyme released plus intracellular enzyme).

2.6 Light microscopy and morphometry

Mast cells were fixed in 2% glutaraldehyde for 2 h. Then, the suspended cells were stained with toluidine blue (0.1% w/v, pH 3.0), put between slides and cover slides, and analyzed under a Nikon Optiphot 2 microscope. Using a magnification of 400×, percentage of degranulated mast cells was quantified. Mast cell was considered active due to the presence of extruded granules near the surface of the cell in question or a toluidine blue stain in half or less of the cell.

2.7 Transmission electron microscopy

Karnovsky's fixative (2% formaldehyde, freshly prepared from paraformaldehyde, 2.5% glutaraldehyde, 0.025% $CaCl_2$, 0.1 M cacodylate buffer, pH 7.4) was used for mast cell fixation. After 1 h in the fixative at 20°C, mast cells were rinsed in 0.2 M cacodylate buffer and postfixed in 1% OsO_4 in 0.1 M cacodylate buffer for at least 2 h at room temperature and dehydrated in ethanol. Next, suspended cells were embedded in Spurr (Pelco, USA). An automatic ultramicrotome (Leica Ultracut R, Austria) was used to cut semithin transverse sections (1 μm), which were stained with filtered 1% toluidine blue. Ultrathin sections (60 nm) were cut with diamond knives, stained with uranyl acetate and lead citrate, mounted on grids (Pelco, USA), and examined in a transmission electron microscope (Zeiss EM 902, Germany).

2.8 Statistical analysis

Results obtained from biochemical and morphometric analyses are presented as ±SEM. Variance analysis was used to determine differences between groups, followed by Tukey-Kramer multiple comparisons test. $P < 0.05$ was considered statistically significant.

3. Results and discussion

Figure 2 shows the effect of varying concentrations of hydroxytyrosol and oleuropein on the mast cells β-hexosaminidase release, induced by concanavalin A, compound 48/80, or calcium ionophore A23187. β-Hexosaminidase release was significantly increased by incubating mast cells with 200 μg/ml concanavalin A or 10 μg/ml compound 48/80 or 50 μg/ml calcium ionophore A23187 solutions, as compared with the corresponding value from the basal group (basal release was always less than 4%). The concanavalin A-induced effects were inhibited by preincubation of mast cells with hydroxytyrosol (10, 50 and 100 μM) or oleuropein (100 μM). Mast cell activation induced by compound 48/80 was inhibited by hydroxytyrosol (100 μM) and oleuropein (100 μM). The calcium ionophore A23187-induced effects were inhibited by preincubation of mast cells with hydroxytyrosol (100 μM) and oleuropein (10, 50 and 100 μM). The inhibitory action of the polyphenols was not accompanied by changes in cell viability (the trypan blue exclusion test indicated a viability of greater than 80%), except for polyphenol concentrations higher than 100 μM (data not shown).

Figure 3 shows a dose-response comparative study with sodium cromoglycate, a mast cell stabilizer, which was used to evaluate the potency of hydroxytyrosol and oleuropein. The inhibitory effect of hydroxytyrosol was higher than that obtained with sodium cromoglycate at the same concentration (100 μM) when mast cells were challenged with concanavalin A. No significant differences were observed when mast cells were challenged with compound 48/80. The oleuropein inhibitory effect was higher than the one obtained with sodium cromoglycate at the same concentrations (10, 50 and 100 μM) when the calcium ionophore A23187 was used to challenge mast cells.

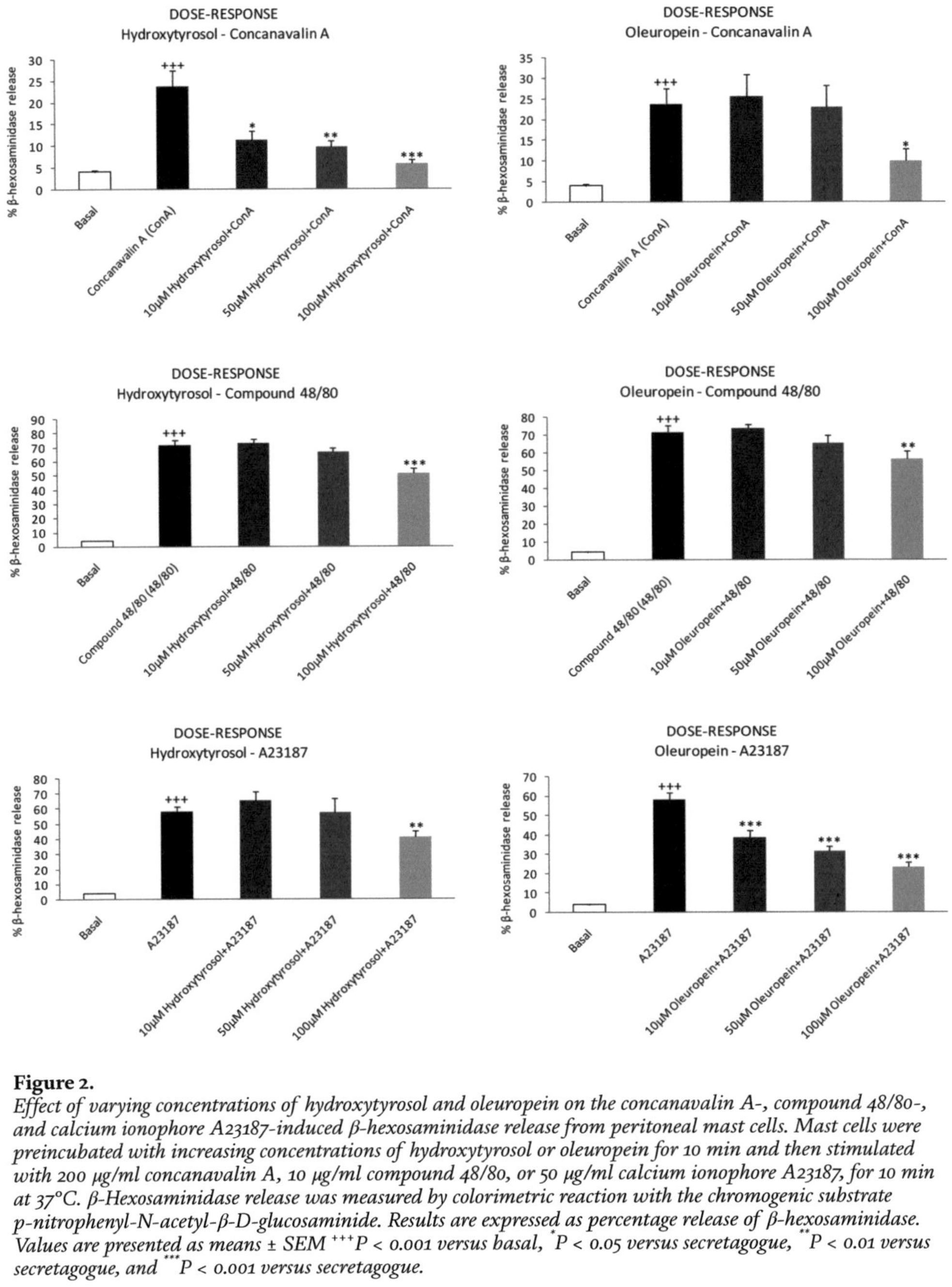

Figure 2.
Effect of varying concentrations of hydroxytyrosol and oleuropein on the concanavalin A-, compound 48/80-, and calcium ionophore A23187-induced β-hexosaminidase release from peritoneal mast cells. Mast cells were preincubated with increasing concentrations of hydroxytyrosol or oleuropein for 10 min and then stimulated with 200 μg/ml concanavalin A, 10 μg/ml compound 48/80, or 50 μg/ml calcium ionophore A23187, for 10 min at 37°C. β-Hexosaminidase release was measured by colorimetric reaction with the chromogenic substrate p-nitrophenyl-N-acetyl-β-D-glucosaminide. Results are expressed as percentage release of β-hexosaminidase. Values are presented as means ± SEM $^{+++}P < 0.001$ *versus basal,* $^{*}P < 0.05$ *versus secretagogue,* $^{**}P < 0.01$ *versus secretagogue, and* $^{***}P < 0.001$ *versus secretagogue.*

The kinetic study results connected with the effect of hydroxytyrosol and oleuropein on β-hexosaminidase release from mast cell are described in **Figure 4**.

The concanavalin A-induced effect was inhibited by hydroxytyrosol (10 min), oleuropein (10 min), and sodium cromoglycate (10 min). Mast cell activation induced by compound 48/80 was only inhibited by sodium cromoglycate (5, 10, and 20 min). The ionophore A23187-induced effect was inhibited by hydroxytyrosol (10 min), oleuropein (5, 10, and 20 min), and sodium cromoglycate (10 and 20 min). The inhibitory action of the test compounds was not accompanied by changes in cell viability (the trypan blue exclusion test indicated a viability of greater than 80%), except for incubation times higher than 20 min (data not shown).

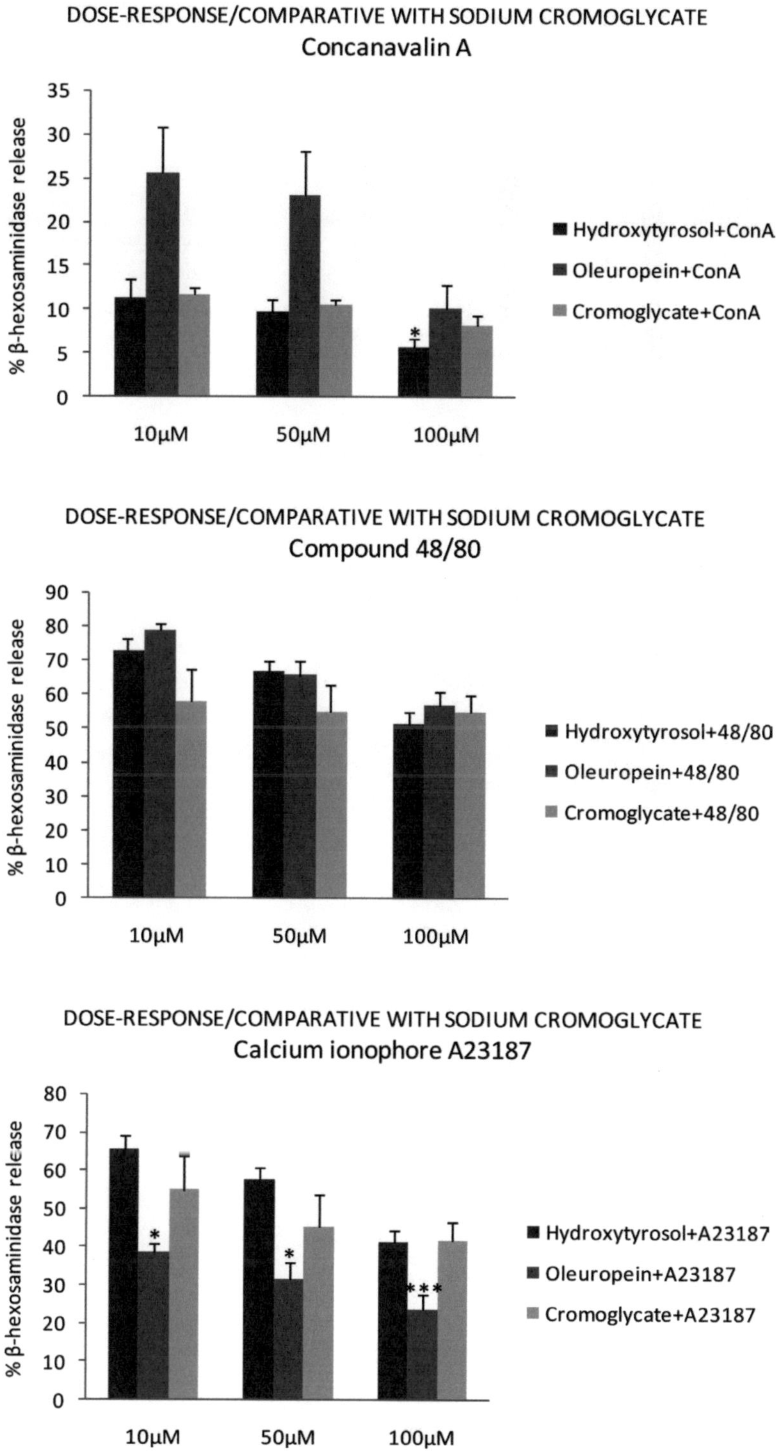

Figure 3.
*Effect of hydroxytyrosol and oleuropein on β-hexosaminidase release from rat peritoneal mast cells, compared with a reference compound, the mast cell stabilizer sodium cromoglycate. Purified mast cells were preincubated with increasing concentrations of hydroxytyrosol, oleuropein, or sodium cromoglycate for 10 min and stimulated with 200 μg/ml concanavalin A, 10 μg/ml compound 48/80, or 50 μg/ml calcium ionophore A23187 for another 10 min at 37°C. β-Hexosaminidase release was measured by colorimetric reaction with the chromogenic substrate p-nitrophenyl-N-acetyl-β-D-glucosaminide. Results are expressed as percentage release of β-hexosaminidase. Values are presented as means ± SEM *P < 0.05 versus sodium cromoglycate and ***P < 0.001 versus sodium cromoglycate.*

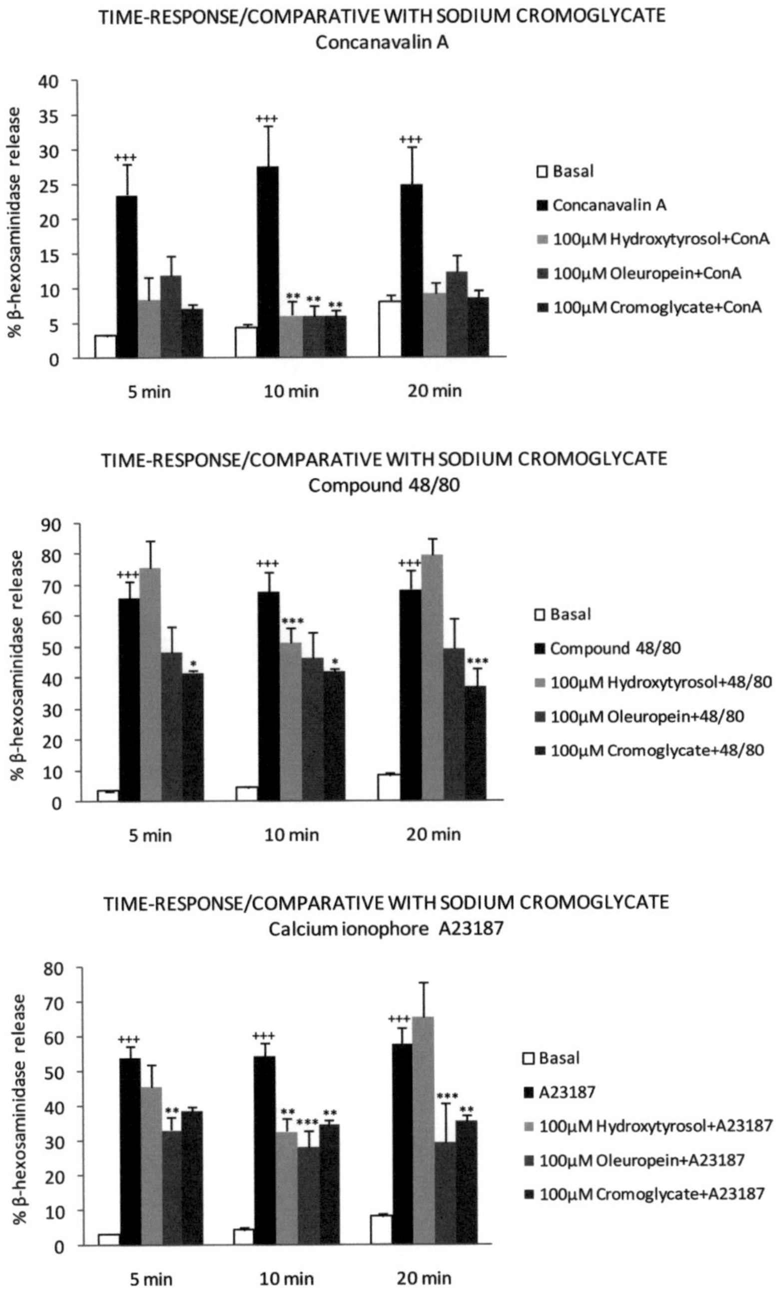

Figure 4.
Kinetic study related to the effect of hydroxytyrosol or oleuropein preincubation on mast cell β-hexosaminidase release, induced by concanavalin A, compound 48/80, or calcium ionophore A23187. Peritoneal mast cells were preincubated for increasing times with 100 μM hydroxytyrosol or oleuropein and then stimulated with 200 μg/ml concanavalin A, 10 μg/ml compound 48/80, or 50 μg/ml calcium ionophore A23187 for 10 min at 37°C. β-Hexosaminidase release was measured by colorimetric reaction with the chromogenic substrate p-nitrophenyl-N-acetyl-β-D-glucosaminide. Results are expressed as percentage release of β-hexosaminidase. Values are presented as means ± SEM $^{+++}P < 0.001$ *versus basal,* $^{*}P < 0.05$ *versus secretagogue,* $^{**}P < 0.01$ *versus secretagogue, and* $^{***}P < 0.001$ *versus secretagogue.*

As a matter of interest, the main findings of this study evidenced that hydroxytyrosol and oleuropein, at non-cytotoxic concentrations, inhibit β-hexosaminidase release from peritoneal mast cells stimulated by different triggers, acting then as mast cell stabilizers.

Mast cell activation is inhibited by several dietary polyphenols, as shown in other in vitro studies [11, 15]. A well-studied immunological effect of polyphenols such as quercetin is their inhibitory action on degranulation, particularly histamine release from mast cells [11, 15, 22]. Quercetin is able to regulate the entry of calcium into mast cells; also the role of apple and grape quercetin and polyphenols in inhibiting mast cell degranulation has been thoroughly documented using in vitro systems such as the RBL-2H3 assay [11, 15]. Epigallocatechin gallate was found to be the active ingredient in green tea extracts for protection against cutaneous inflammation. This compound may also inhibit mast cell histamine release stimulated with both a calcium ionophore and an IgE-antigen complex [23]. Kaempferol, myricetin, phloretin, and luteolin also proved to be effective inhibitors of histamine release [11, 15]. However, hydroxytyrosol and oleuropein were never studied on mast cell mediator release in response to either immune or nonimmune triggers.

In addition, we have also proven that the inhibitory effects by hydroxytyrosol were stronger than those of oleuropein or the reference compound sodium cromoglycate when mast cells were activated by concanavalin A. On the contrary, oleuropein resulted to be a more efficient inhibitor of mast cell degranulation than hydroxytyrosol or sodium cromoglycate when cells were challenged with the calcium ionophore A23187. **Table 1** presents EC_{50} values for each compound. These results strongly suggest that the inhibitory activity might be defined by the nature of the stimulus for β-hexosaminidase release and the chemical structure of both polyphenols. Each of the secretagogues takes a different mast cell activation pathway, and these pathways may be differentially sensitive to the action of hydroxytyrosol or oleuropein. Mast cells may be activated via several different mechanisms, among which is the classical pathway known as immunological or IgE-mediated mast cell activation, triggered by the cross-linking of FcεRI receptors. Mast cell activation may also be completed in an IgE-independent manner using commercially available activators, such as basic secretagogues and calcium ionophores. Concanavalin A is a glucose-/mannose-specific lectin that activates mast cells by a mechanism similar to the antigen-antibody reaction. This lectin causes FcεRI receptors to cluster on cell membranes. In turn, this triggers a series of intracellular events leading to the secretion of mast cell mediators by exocytosis. Phosphatidylserine markedly enhances the release of mediators from rat mast cells by concanavalin A [24–26]. The synthetic compound 48/80, a basic secretagogue, activates heterotrimeric G proteins by enhancing the dissociation of GDP from $G\alpha$ subunits, thus accelerating the event considered as a rate-limiting step in conventional G protein activation by receptors coupled to heterotrimeric G proteins [27]. Calcium ionophore A23187, a mobile carrier of divalent cations such as Ca^{2+}, Mg^{2+}, and double H^{+}, may reduce the level of calcium stored in mitochondria or increase the inflow from the extracellular medium, resulting in a cytosolic calcium increase, which may induce mast cell exocytosis and preformed mediator release. Calcium release from internal stores has been shown to be related to some second messengers, including phospholipase C, phospholipase D, inositol

	Hydroxytyrosol	Oleuropein	Sodium cromoglycate
Concanavalin A	7.58 ± 4.9	67 ± 5.1	95 ± 8.9
Compound 48/80	61 ± 5.0	59 ± 5.0	97 ± 9.1
Calcium ionophore A23187	64 ± 5.0	22 ± 1.8	54 ± 5.1

Table 1.
EC_{50} (μM) ± SEM values for inhibitory activity of hydroxytyrosol, oleuropein, and sodium cromoglycate on concanavalin A-, compound 48/80-, and calcium ionophore A23187-induced mast cell activation.

1,4,5-triphosphate, and diacylglycerol. Degranulation dependent on the influx of extracellular calcium may be associated to the members of the SNARE (soluble NSF attachment protein receptor) family, such as synaptosome-associated protein of 23 kDa (SNAP-23), syntaxin, synaptotagmin, and molecules of the vesicle-associated membrane protein (VAMP) family which regulate the granule-to-granule or granule-to-plasma membrane fusion process [28]. Sodium cromoglycate is a compound commonly used in the treatment of allergic diseases. The effect of sodium cromoglycate is due to its ability to stabilize the mast cell membrane and to prevent the release of histamine and inflammatory mediators [29]. Mechanisms of the action of sodium cromoglycate include blocking of the influx of calcium into mast cells, inhibition of phosphodiesterase, and regulation of phosphorylation of mast cell proteins [30].

The findings of this study suggest that hydroxytyrosol and oleuropein act at different molecular sites. It seems that the hydroxytyrosol inhibitory mechanism may be related, at least partially, to inhibition of the cross-linking of high-affinity receptors for IgE (FcεRI) or to a probable interaction of the polyphenol with concanavalin A. Reports show that polyphenols form soluble and insoluble complexes with proteins and may cause them to become hypoallergenic by either modifying the structure of the allergenic protein or making it less bioavailable [15]. Moreover, two mechanisms are proposed for the mast cell inhibitory action of quercetin and other polyphenols: one where polyphenols impact allergen-IgE complex formation and another one where the polyphenols impact on the complex binding to their receptor (FcεRI) on mast cells [15].

Our results also suggest that oleuropein seems to block signaling pathways downstream of cytosolic calcium increase. However, further research is needed in order to explain the exact molecular mechanisms of these actions.

Despite the strong biochemical evidence, we considered a morphological evaluation necessary to reinforce the validity of our initial findings. Thus, a second set of experiments was designed to analyze the effect of the hydroxytyrosol and oleuropein on mast cell morphology by light and electron microscopy.

Peritoneal mast cells were easily identified by the presence of a cytoplasm dominated by distinctive secretory granules which stain metachromatically (**Figure 5A–J**). A representative mast cell from the basal group is shown in **Figure 5A**. Tightly packed secretory granules dominate the cytoplasm. Characteristic mast cells stimulated with concanavalin A, compound 48/80, and calcium ionophore A23187 are shown in **Figure 5B–D**, respectively. Cell surface disruption, typical of degranulating mast cells, may be observed. **Figure 5E–G** show mast cells treated with 100 μM hydroxytyrosol, for 10 min, prior to the challenge with concanavalin A, compound 48/80, and calcium ionophore A23187, respectively. **Figure 5H–J** show mast cells treated with 100 μM oleuropein, for 10 min, prior to the challenge with concanavalin A, compound 48/80, and calcium ionophore A23187, respectively. The morphology of the cells treated with the polyphenols shows a lower degree of degranulation than that of secretagogue samples.

Figure 6A–J shows mast cells observed under transmission electron microscope. **Figure 6A** shows a characteristic view of the basal mast cell population. These cells are characterized by a non-segmented, irregular nucleus with only moderate nuclear chromatin condensation, narrow surface folds, and numerous secretory granules regularly distributed throughout the cell cytoplasm. Almost all granules display either round or oval profiles and appear homogenously dense. Representative mast cells from the concanavalin A-, compound 48/80-, and calcium ionophore A23187-treated group are shown in **Figure 6B–D**, respectively.

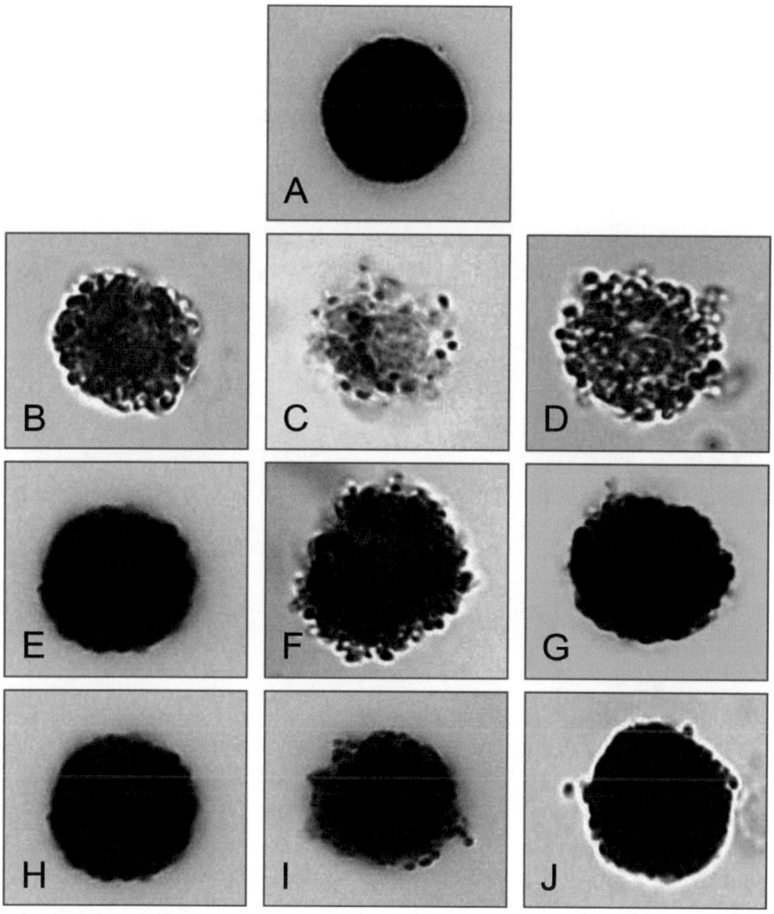

Figure 5.
Light microscopic photographs of peritoneal mast cells (toluidine blue stain). Mast cells were preincubated with 100 μM hydroxytyrosol or oleuropein for 10 min and then stimulated with 200 μg/ml concanavalin A, 10 μg/ml compound 48/80, or 50 μg/ml calcium ionophore A23187, for 10 min at 37°C. After incubations, cells were fixed and stained for light microscopy. (A) Basal. The cytoplasm is dominated by closely packed secretory granules. (B–D) Concanavalin A (ConA), compound 48/80 (48/80), and calcium ionophore A23187 (A23187), respectively. Degranulating cells may be seen. (E–G) hydroxytyrosol + ConA, hydroxytyrosol + 48/80, and hydroxytyrosol + A23187, respectively. (H–J) Oleuropein + ConA, oleuropein + 48/80, and oleuropein + A23187, respectively. The morphology of the polyphenol-treated cells shows a lower degree of degranulation than that of secretagogue samples. 600×.

These cells display obvious morphological changes and show evidence of increased granule release by exocytosis compared to basal cells. Mast cells exhibit a cytoplasm with irregular secretory granules showing various degrees of electron densities. Perigranular dilated and electron-lucent spaces surround some granules. Some of these spaces look fused, forming multiple cavities and intracytoplasmic channels. **Figure 6E–G** show mast cells treated with 100 μM hydroxytyrosol, for 10 min, prior to the challenge with concanavalin A, compound 48/80, and calcium ionophore A23187, respectively. **Figure 6H–J** show mast cells treated with 100 μM oleuropein, for 10 min, prior to the challenge with concanavalin A, compound 48/80, and calcium ionophore A23187, respectively. The morphology of the polyphenol-treated cells shows a lower degree of degranulation than that of secretagogue samples.

Our biochemical results about β-hexosaminidase release are consistent with those obtained by light and electron microscopy.

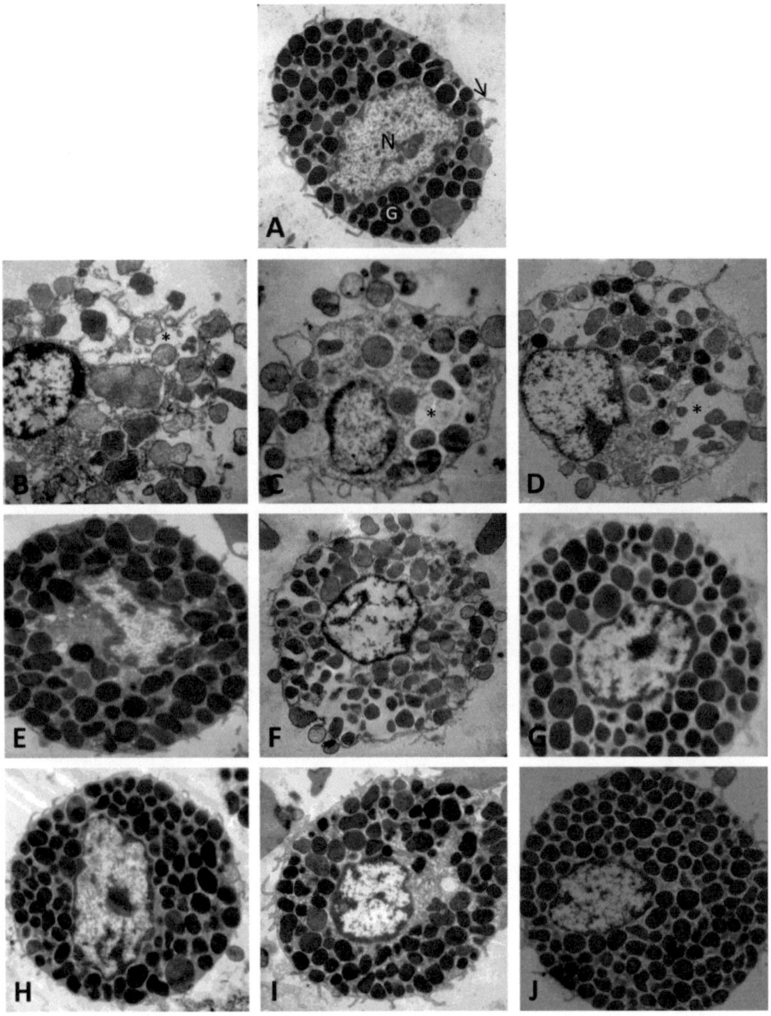

Figure 6.
Transmission electron micrographs showing peritoneal mast cells. Mast cells were preincubated with 100 μM hydroxytyrosol or oleuropein for 10 min and then stimulated with 200 μg/ml concanavalin A, 10 μg/ml compound 48/80, or 50 μg/ml calcium ionophore A23187, for 10 min at 37°C. After incubations, cells were processed for electron microscopy. (A) Basal. A non-segmented nucleus (N), narrow surface folds (red arrows), and numerous secretory granules (G) distributed throughout the cell cytoplasm are observed. (B–D) Concanavalin A (ConA), compound 48/80 (48/80), and calcium ionophore A23187 (A23187), respectively. Granules surrounded by perigranular dilated and electron-lucent spaces are evident. Some granules are seen out of mast cell cytoplasm, and others appear fused, forming cavities and intracytoplasmic channels (asterisk). (E–G) Hydroxytyrosol + ConA, hydroxytyrosol + 48/80, and hydroxytyrosol + A23187, respectively. (H–J) Oleuropein + ConA, oleuropein + 48/80, and oleuropein + A23187, respectively. Polyphenol-treated cells show minimal degranulation. 5000×.

4. Conclusions

In conclusion, our present findings reveal for the first time that hydroxytyrosol and oleuropein, the major phenolic compounds in olive, inhibit mast cell degranulation triggered by both immune and nonimmune pathways. Our discoveries also suggest that olive polyphenols, especially hydroxytyrosol and oleuropein, may provide insights to develop useful tools to prevent and treat mast cell-mediated disorders. These findings may be of interest to the immunopharmacology industry or could even lead in determining the ideal concentrations of each component in virgin olive oils to be able to label them as healthy oils. However, further tests are needed in order to generate hypoallergenic products via polyphenol treatment.

Acknowledgements

I thankfully acknowledge Dr. Fabio A. Persia, Dra. María Laura Mariani, Ing. Norberto Domizio, and Ing. Elisa Bocanegra for their technical support. Funding for this work was provided by grants from the Consejo Nacional de Investigaciones Científicas y Técnicas (PIP-CONICET 5128) and Secretaría de Ciencia, Técnica y Postgrado de la Universidad Nacional de Cuyo (Research Proyects SeCTyP UNCuyo 06/J268-06/J395 and Research Program CS-453-2010), Argentina.

Conflict of interest

The author declares no conflict of interest.

Abbreviations

ConA	concanavalin A
48/80	compound 48/80
A23187	calcium ionophore A23187
FcεRI	high-affinity receptor for IgE
IgE	immunoglobulin E

Author details

Alicia Beatriz Penissi[1,2]

1 Facultad de Ciencias Médicas, Instituto de Histología y Embriología "Dr. Mario H. Burgos" (IHEM-CCT Mendoza-CONICET), Universidad Nacional de Cuyo, Argentina

2 Facultad de Ciencias Médicas, Instituto de Investigaciones, Universidad del Aconcagua - Mendoza, Argentina

*Address all correspondence to: apenissi@yahoo.com.ar

References

[1] Bischoff S. Role of mast cells in allergic and non-allergic immune responses: Comparison of human and murine data. Nature Immunology. 2007;7:93-104

[2] Galli S, Borregaard N, Wynn T. Phenotypic and functional plasticity of cells of innate immunity: Macrophages, mast cells and neutrophils. Nature Immunology. 2011;**12**:1035-1044

[3] Sur R, Cavender D, Malaviya R. Different approaches to study mast cell functions. International Immunopharmacology. 2007;7:555-567

[4] Kumar V, Sharma A. Mast cells: Emerging sentinel innate immune cells with diverse role in immunity. Molecular Immunology. 2010;**48**:14-25

[5] Kraneveld A, Sagar S, Garssen J, Folkerts G. The two faces of mast cells in food allergy and allergic asthma: The possible concept of Yin Yang. Biochimica et Biophysica Acta. 2011;**1822**:93-99

[6] Weller C, Collington S, Williams T, Lamb J. Mast cells in health and disease. Clinical Science. 2011;**120**:473-484

[7] Theoharides T, Alysandratos K, Angelidou A, Delivanis D, Sismanopoulos N, Zhang B, et al. Mast cells and inflammation. Biochimica et Biophysica Acta. 2012;**1822**:21-33

[8] Yamada P, Zarrouk M, Kawasaki K, Isoda H. Inhibitory effect of various Tunisian olive oils on chemical mediator release and cytokine production by basophilic cells. Journal of Ethnopharmacology. 2008;**116**:279-287

[9] Tore F, Tuncel N. Mast cells: Target and source of neuropeptides. Current Pharmaceutical Design. 2009;**15**:3433-3445

[10] Kalesnikoff J, Galli S. New developments in mast cell biology. Nature Immunology. 2008;**9**:1215-1223

[11] Middleton E, Kandaswami C, Theoharides C. The effects of plant flavonoids on mammalian cells: Implications for inflammation, heart disease, and cancer. Pharmacological Reviews. 2000;**52**:673-751

[12] Penissi A, Vera M, Mariani L, Rudolph M, Ceñal P, de Rosas C, et al. Novel anti-ulcer α,β-unsaturated lactones inhibit compound 48/80-induced mast cell degranulation. European Journal of Pharmacology. 2009;**612**:122-130

[13] Sakai S, Sugawara T, Matsubara K, Hirata T. Inhibitory effect of carotenoids on the degranulation of mast cells via suppression of antigen-induced aggregation of high affinity IgE receptor. The Journal of Biological Chemistry. 2009;**284**:28172-28179

[14] Vera M, Persia A, Mariani L, Fogal T, Ceñal P, Tonn C, et al. Activation of the human mast cell line LAD2 is modulated by dehydroleucodine and xanthatin. Leukemia Lymphoma. 2012;**53**:1795-1803

[15] Singh A, Holvoet S, Mercenier A. Dietary polyphenols in the prevention and treatment of allergic diseases. Clinical and Experimental Allergy. 2011;**41**:1346-1359

[16] Tripoli A, Giammanco M, Tabacchi G, Di Majo D, Giammanco S, La Guardia M. The phenolic compounds of olive oil: Structure, biological activity and beneficial effects on human health. Nutrition Research Reviews. 2005;**18**:98-112

[17] Jemai H, Bouaziz M, Fki I, El Feki A, Sayadi S. Hypolipidimic and antioxidant activities of oleuropein and

its hydrolysis derivative-rich extracts from Chemlali olive leaves. Chemico-Biological Interactions. 2008;**176**:88-98

[18] Hagiwara K, Goto T, Araki M, Miyazaki H, Hagiwara H. Olive polyphenol hydroxytyrosol prevents bone loss. European Journal of Pharmacology. 2011;**662**:78-84

[19] Mousli M, Bronner C, Bueb JL, Tschirhart E, Gies JP, Landry Y. Activation of rat peritoneal mast cells by substance P and mastoparan. The Journal of Pharmacology and Experimental Therapeutics. 1989;**250**:329-335

[20] MacGlashan D, Guo C. Oscillations in free cytosolic calcium during IgE-mediated stimulation distinguish human basophils from human mast cells. Journal of Immunology. 1991;**147**:2259-2263

[21] Puri N, Roche P. Mast cells possess distinct secretory granule subsets whose exocytosis is regulated by different SNARE isoforms. Proceedings of the National Academy of Sciences of the United States of America. 2008;**105**:2580-2585

[22] Weng Z, Zhang B, Asadi S, Sismanopoulos N, Butcher A, Fu X, et al. Quercetin is more effective than cromolyn in blocking human mast cell cytokine release and inhibits contact dermatitis and photosensitivity in humans. PLoS One. 2012;**7**:1-10

[23] Maeda-Yamamoto M, Ema K, Monobe M, Tokuda Y, Tachibana H. Epicatechin-3-O-(3″-O-methyl)-gallate content in various tea cultivars (*Camellia sinensis* L.) and its in vitro inhibitory effect on histamine release. Journal of Agricultural and Food Chemistry. 2012;**60**:2165-2170

[24] Hirata F, Axelrod J, Crews F. Concanavalin A stimulates phospholipid methylation and phosphatidylserine decarboxylation in rat mast cells. Proceedings of the National Academy of Sciences of the United States of America. 1979;**76**:4813-4816

[25] Hosono H, Aoki J, Nagai Y, Bandoh K, Ishida M, Taguchi R, et al. Phosphatidylserine-specific phospholipase A1 stimulates histamine release from rat peritoneal mast cells through production of 2-acyl-1-lysophosphatidylserine. The Journal of Biological Chemistry. 2001;**276**:29664-29670

[26] Lopes F, Cavada S, Pinto P, Sampaio H, Gomes C. Differential effect of plant lectins on mast cells of different origins. Brazilian Journal of Medical and Biological Research. 2005;**38**:935-941

[27] Palomäki V, Laitinen T. The basic secretagogue compound 48/80 activates G proteins indirectly via stimulation of phospholipase D-lysophosphatidic acid receptor axis and 5-HT1A receptors in rat brain sections. British Journal of Pharmacology. 2006;**147**:596-606

[28] Wang H, Wang HS, Liu P. Agents that induce pseudo-allergic reaction. Drug Discoveries & Therapeutics. 2011;**5**:211-219

[29] Shin H, Kim J, An H, Park K, Kim M. Effect of disodium cromoglycate on mast cell-mediated immediate-type allergic reactions. Life Sciences. 2004;**74**:2877-2887

[30] Nishibori M, Saeki K. Disodium cromoglycate inhibition of substance P-induced histamine secretion is calcium dependent. Japanese Journal of Pharmacology. 1983;**33**:1255-1261